OBSERVER'S GUIDE TO
HALLEY'S COMET

JAMES MUIRDEN

GEORGE PHILIP

Also by James Muirden
Astronomy with a Small Telescope

Published by George Philip,
12-14 Long Acre, London WC2E 9LP

First published by Arco Publishing, Inc.,
215 Park Avenue South, New York, NY 10003

British Library Cataloguing in Publication Data
Muirden, James
    Observer's guide to Halley's Comet
    1. halley comet
    I. title
    523.6'4   QB723.H2
    ISBN 0 540 01095 2

Printed in Great Britain by
J. W. Arrowsmith Ltd., Bristol

Contents

Preface

I hope that this book will help you to see and enjoy the spectacle of Halley's comet in 1985 and 1986.

Halley's comet, as seen from our northern latitudes, will *not* be a great spectacle. Yet everybody, understandably, will want to catch a sight of it. This guide is intended to help someone with no prior astronomical knowledge of any sort. The only other helpful accessory will be a pair of binoculars or a telescope.

If you use this guide carefully, you have a good chance of following the comet over a period of several months, rather than just over the few weeks when it attracts the most attention. You will see it brighten and dim again as it sets off once more on its immense 76-year orbit that covers 12 billion kilometers. If you have a camera, you may even manage to photograph it!

If you live in the Earth's southern hemisphere, then your chances of enjoying a good view are much greater, since the comet should become easily visible to the naked eye in March and April of 1986. However, the close-up maps will still be useful if you wish to identify it during the fainter stages, and you can locate these map positions using a planisphere.

I must thank Nigel Code for one of the illustrations; the Council of the British Astronomical Association, for permission to quote from one of their publications; Henry Rasof, for suggesting that I write this book in the first place; and my son Daniel, for help with computing.

JAMES MUIRDEN

1

What Is a Comet?

You are reading this book because you wish to find out something about Halley's comet. You have heard of comets, and of Halley's in particular. But you may not have the faintest idea of what a comet *is*, apart, perhaps, from the fact that it shines in the sky, has a long tail, and may make some folk wonder if the end of the world is nigh!

But this is a bad start, for most comets are devoid of a noticeable tail anyway. Furthermore, there are so many of them flying around through space that the world must come to its termination at least a dozen times a year, for comets are seen about as frequently as this.

Comets still take astronomers by surprise, for they can come into view unexpectedly. But this is not because they suddenly manifest themselves out of nothing. They are at least as old as Earth; but they are normally too far away in space to be made out. Only when the path, or *orbit*, of a comet brings it near the vicinity of Earth and Sun is it likely to be detected as a hazy glow in the night sky. A "new" comet is simply one that is approaching the telescopes of astronomers for the first recorded time.

Our Earth, and the other eight planets that make up the solar system, all revolve around the Sun, held at the end of an invisible rope of gravity. This is the same force that dominates the comets—other members of the Sun's scattered family. However, there are two important differences between planets and comets:

1. Whereas Earth and the other eight major planets have diameters

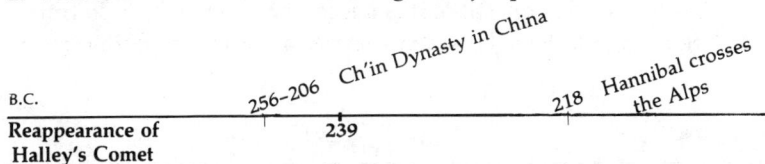

B.C.

256–206 Ch'in Dynasty in China

218 Hannibal crosses the Alps

Reappearance of Halley's Comet

239

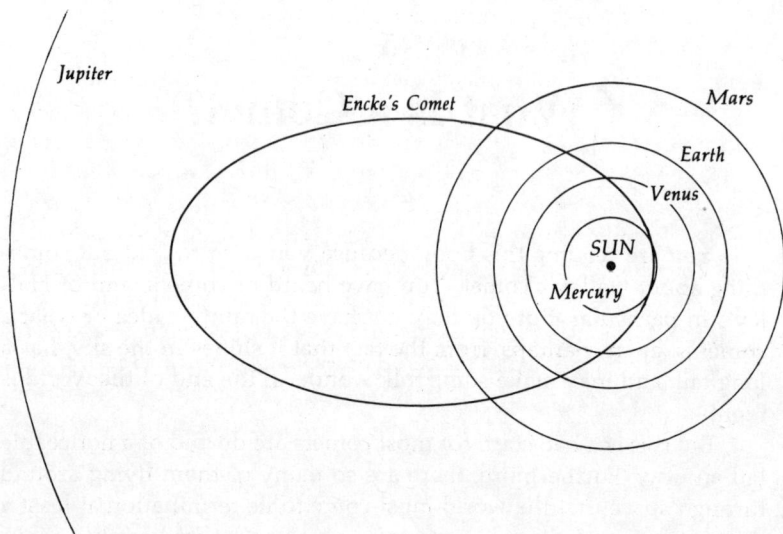

Encke's comet has the shortest period of any known comet—only three years and four months. This diagram shows its path in relation to the orbits of the planets from Mercury to Jupiter.

to be measured in thousands of kilometers, the solid part of a comet is only a few kilometers across.

2. The orbits of most comets are much more elongated, or *eccentric*, than are those of the planets. The diagram represents the orbit of a well-known comet (Encke's comet) in correct relationship with the orbits of the planets from Mercury to Jupiter.

What Is a Comet Made Of?

To answer this question, it is necessary to think back in time to the beginnings of the solar system, some four and a half billion years

215 Building of the Great Wall of China

207–A.D. 221 Han Dynasty in China

196 Rosetta Stone carved

ago. Astronomers believe that the whole family of Sun and planets condensed out of a vast cloud of *gust* (the writer's suggested name for the mixture of gas and "dust" that this cloud must have contained). A sample of gust would have contained all the known elements of the universe—92 of them—in the form either of separate dancing molecules (gas) or clumps of molecules forming minute solid fragments (dust). By far the most common constituent was the gas hydrogen, but other gases such as oxygen and nitrogen also abounded; while the dust grains contained iron, carbon, silicon, and many other elements familiar to us in their solid rather than gaseous state.

When this huge soup of material began to condense into solid bodies, a large proportion gathered at the center of the cloud, pulled there by the effect of gravity. It was a kind of "runaway" attraction: the greater the amount of material accumulating at the center, the stronger its pull and appetite for more became. It couldn't go on growing indefinitely, however, since the supply began to give out. There was another reason, too: *thermonuclear reactions* started at the center of the infant Sun, inside this compacting mass of gust where the temperature began rising dramatically. It reached millions of degrees, stripping hydrogen atoms into their subatomic components and rebuilding them into a more stable element, helium. Once formed, helium cannot be cracked asunder; but in the process of its formation great quantities of energy are released. The Sun became self-sustaining, the immense force of energy being generated at its center both keeping the nuclear reactions going and counteracting the tendency for it to shrink into a much smaller body.

The Sun is believed to have been shining at its present power, or thereabouts, for billions of years, and the signs are that it can continue to do so for at least as long into the future. What of the planets and comets? The planets, evidently, were mini-condensations within the gust cloud. They grew hot as they collapsed into spherical bodies, but not hot enough to become hydrogen bombs like the Sun. However, the heat was sufficient to melt the solid dust into liquid metal and rock, which then cooled into recognizable worlds of vari-

146 Carthage destroyed by Scipio

120 Building of the Temple of Apollo at Delphi

102 Birth of Julius Caesar (d. 44)

ous sizes, some larger and some smaller than Earth.

It seems certain that the comets did not become hot enough even to melt into rock-and-metal blobs. They were such tiny condensations, compared with the planets, that they did not become particularly hot at all. Therefore, their material is crumbly rather than solid. This lack of heating also means that they did not boil off all the water that formed when hydrogen and oxygen atoms met, but kept it as ice. In fact, it is possible that a comet is like a huge block of ice with dust grains embedded in it—this is the *dirty snowball* model developed by astronomer Fred Whipple.

Why Do Comets Brighten and Fade?

When brought too near a fire, a snowball melts. The Sun is a fierce fire indeed (surface temperature 6000° C, some three times as hot as a steel furnace), and an icy comet approaching its vicinity begins to suffer the same fate. However, instead of melting, the ice *sublimes,* turning directly into a gas. This happens because the material is exposed to the vacuum of interplanetary space. Liquid water cannot exist in a vacuum; it requires the pressure of an atmosphere, such as our own, to preserve it in this state, without which the separate molecules fly apart to form a gaseous cloud. Therefore, as the cometary body warms up, a mist begins to surround it; and this misty cloud can extend through a huge volume of space, thousands or millions of times greater than that of the solid part, or *nucleus,* at the center.

This misty cloud is called the *coma.* It contains dust as well as gas, for the dust particles around the surface of the nucleus are released into space when the ice sublimes. At the center of the coma is the *central condensation,* a brighter region that may be very small and almost starlike, or so vague that it is barely distinguishable from the coma itself. People often refer to this as the "nucleus," but they are

73   The Slave War of Spartacus   30   Suicide of Antony and   20   Herod begins rebuilding
                                                  Cleopatra   of the Temple

11

wrong; the true nucleus is too small to be seen in any but the largest telescopes. Active comets may throw off distinct shells, or *envelopes*, of material from the nucleus, and these can appear as bright rays or fragments surrounding the coma. The whole system of nucleus, condensation, coma, and envelopes (if any) constitutes the *head* of a comet.

A *tail* does not become conspicuous unless the comet passes fairly near the Sun—probably within the orbit of Earth—and has large quantities of gas and dust being released from its nucleus. The tail points more or less away from the Sun, since the "wind" of atomic particles flying outward from our star in the fury of its radiation blows the cometary material away with it. But do not imagine that this gale compares in any way with the winds of our experience, for the density of a comet's tail is only a millionth of the density of air, and the only reason for its visibility is its enormous bulk: remember that the material expelled from the surface of a nucleus only a few kilometers across may extend for millions of kilometers through space!

"Young" and "Old" Comets

It seems a reasonable guess that the cometary bodies were formed at the same time as the planets, and therefore are the same age—about four and a half billion years. However, inanimate matter does not "age" unless something is done to it. What affects a comet is the frequency of its passages close to the Sun—the point on its orbit known as *perihelion*. It is during perihelion passage that material is boiled and blasted away. If the time taken to orbit the Sun is only a few Earth-years, then the body is likely to have been more stripped of its volatile materials than if it takes thousands of years to go from one perihelion to the next. In the former case the comet may have little or no tail, even when closest to the Sun, and its coma may be

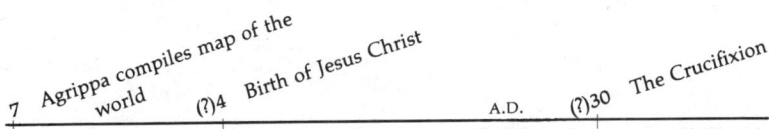

Agrippa compiles map of the world   (?)4   Birth of Jesus Christ   A.D.   (?)30   The Crucifixion

7

small and dim, for most of the ices and solid fragments have been lost to space in previous displays. Astronomers have discovered some small planetlike bodies moving in cometary orbits, but showing no sign of haziness about their sharp discs. Presumably these are the nuclei of ancient comets that may once have blazed gloriously in the sky, but now orbit, dim and dead, through the planetary realm.

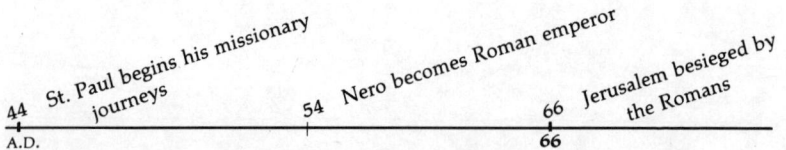

44 St. Paul begins his missionary journeys

54 Nero becomes Roman emperor

66 Jerusalem besieged by the Romans

A.D.

66

2

Some Famous Comets

Of the dozen or 20 comets that pass perihelion each Earth-year, only one or two are likely to become visible without a telescope. However, do not imagine that these are likely to be spectacular objects; at best, they will be a faint smudge to be found somewhere in the patterns of the stars, and no nonastronomers will become very excited, unless they are anxious to be able to tell their friends that they have seen a comet! As night succeeds night, the hazy patch will be found to have moved to different star fields, and eventually it will fade away, its passing acquaintance with Earth ended.

It is a fact that very few of the "periodic" comets—those with an orbital period of less than a couple of centuries, and which have been observed at a previous return and are therefore expected back—are bright enough to be seen with the naked eye at all. Halley's comet is an exception. All the really spectacular objects have had periods of many centuries, or even thousands of years. They have never been seen before during the relatively brief history of precise astronomical recording, and so their reappearance is utterly unexpected.

There are numerous references to cometary appearances in historical records, although they are often vague. A bright comet chronicled in the year 239 B.C. was evidently Halley's, making the first of its 29 recorded appearances. One observed in A.D. 146 was likened to the Sun in brilliance: a pardonable exaggeration, perhaps, if we allow for amazement and fear! During the past thousand years, we find about two dozen comets attracting widespread attention:

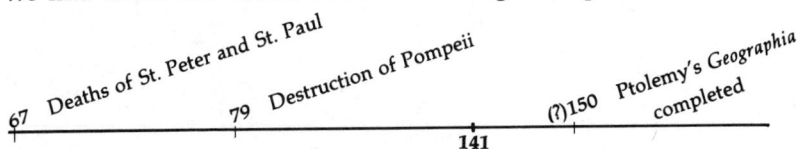

67 Deaths of St. Peter and St. Paul 79 Destruction of Pompeii (?)150 Ptolemy's Geographia completed

141

De Chèseaux's comet of 1744 was a famous object. It had six tails, and is seen here rising before the Sun. (Drawing by the author, based on a contemporary picture.)

some with tails stretching halfway or more across the sky, and others so bright that their heads were detectable in the daylight blue!

This may be a good point at which to mention the relationship of the motion of comets across the sky to the position of the Sun. The Sun is the focus of a comet's flight, and therefore it is usual for a comet to be brightest and most spectacular when it passes near the Sun in the sky. This may bring problems, for the brilliance of the sky around the Sun will drown all but the brightest comets as their daily motion carries them into the sunset or sunrise twilight. When Halley's comet passes perihelion in early February 1986, it will probably be lost from view, in amateur instruments, for the best part of a month.

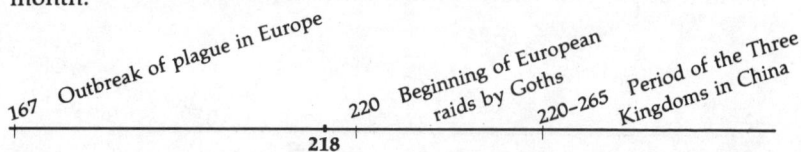

167   Outbreak of plague in Europe   220   Beginning of European   Period of the Three
                                                  raids by Goths   220-265   Kingdoms in China

218

One of the most sensational comets of all time appeared in the year 1264; when seen before dawn, with its bright head in the eastern sky rising before the Sun, its tremendous tail extended more than halfway across the sky. More famous, since it occurred at a time of increasing astronomical activity, is the Great Comet of 1577, with a head as bright as the planet Venus—which can shine so brilliantly in the sky that it has been observed to cast a shadow! Numerous pamphlets and posters were rushed off the presses of Europe, anticipating its astrological significance. It is significant, too, for having been carefully observed by the astronomer Tycho Brahe, who proved that it moved through interplanetary space. Up to that time—and, indeed, for some time after—it was popularly believed that a comet was a fiery phenomenon in Earth's upper atmospheric regions.

The bright comet of 1680 is interesting for two reasons. It be-

The great comet of 1843, as seen from Southern England. (Drawing by the author, based on a contemporary sketch.)

265–420 Tsin Dynasty in China

295 Emperor Diocletian imposes death penalty on Christians

324 Building of Constantinople commences

295

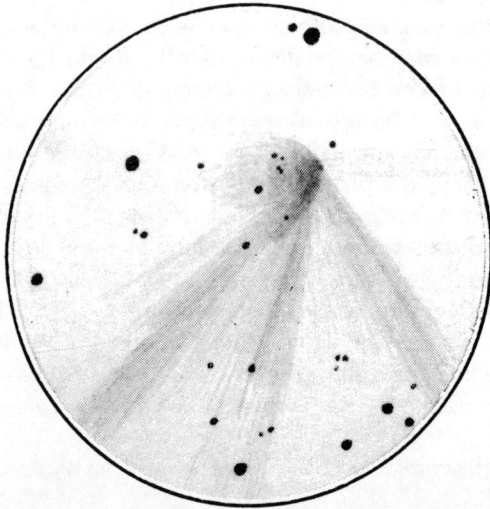

A binocular view of comet IRAS-Araki-Alcock, 1983. It was made by Eric Alcock, one of the independent discoverers.

longs to a group known as "Sun-grazers"—comets whose orbits carry them desperately near the Sun. At perihelion, on December 18, 1680, this comet was only 200,000 kilometers from the solar surface—750 times as close as Earth! Yet at its greatest distance, or *aphelion*, which may not be reached for another thousand years, it will lie some ten times as far away from the Sun as the outermost known planet, Pluto. Studies of the motion of this comet helped Isaac Newton realize that it obeyed the gravitational laws which he was in the process of deriving to describe the movements of the planets—a most important testimony to the "universality" of gravity.

One of the most beautiful comets of all time must have been the one that appeared in 1744, and at its greatest development exhibited a set of six tails when viewed with the naked eye. Another fine one was seen in 1811, and since the French vintage happened to be par-

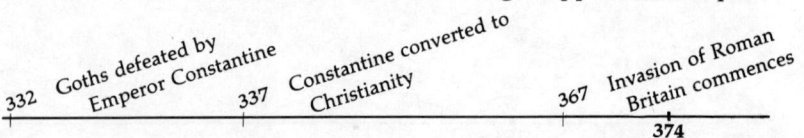

332 Goths defeated by Emperor Constantine

337 Constantine converted to Christianity

367 Invasion of Roman Britain commences

374

ticularly excellent, the produce was dubbed "Comet wine"! The bright comet of 1843, another Sun-grazer, caused so much popular excitement in the United States that it led to the founding of the nation's first observatory, at Harvard. At perihelion, its velocity around the Sun reached 550 kilometers per second. In fact, this superb object was not detected until the day after perihelion (February 28), when it was seen within a few degrees of the Sun, in broad daylight.

The beautiful comet of 1858, Donati's, is famous for its wide, curved tail, although it was not so brilliant as some others. In 1882, another daylight comet was seen. This one was observed to break up into four separate nuclei after perihelion passage, and it will probably return as four separate comets, at intervals of a century, many hundreds of years hence.

Coming to the present time, comet Ikeya-Seki of 1965 was a glorious sight, and so was Bennett's comet of 1970. The brightest comet since World War II was West's comet of 1976, another Sun-grazer, which divided into separate nuclei in the style of the Great Comet of 1882, and was detectable with the naked eye before the Sun had set. However, there was much less public excitement about this object than might be expected, considering its brilliance; in fact, no comet of the present century (with the exception, perhaps, of the Daylight Comet of 1910) has received as much popular attention as the great comets of the past. Very likely, this is due to the relentless spread of permanent nocturnal lighting, which dims the stars drastically compared with their appearance to our forefathers. This consideration will certainly affect the appearance of Halley's comet to most townsfolk.

However, the last few years have seen two interesting cometary appearances. In August 1979, an American satellite making routine solar observations recorded a comet passing so close to the Sun that it may well have collided with the solar surface and been utterly destroyed. No Earth-based observatory recorded it at all, and there has been no previous evidence for a comet having a perihelion position *within* the Sun!

393 Last of the ancient Olympic Games

420–589 Period of the Southern Dynasties in China

446 Rise of Attila the Hun

451

The second curiosity was discovered in May 1983, when the fourth comet to be detected in that year was picked up independently by two Earth-based observers and a satellite—the first and so far only time that such a thing has happened. Moreover, this comet (known as IRAS-Araki-Alcock) turned out to be traveling almost directly toward Earth, and on May 11 it passed by at a distance of only 12 times that of the Moon, or four and a half million kilometers. This may not sound very close, but only a comet observed back in 1770 is known to have given Earth a closer graze. However, it was not spectacular in the sense of being bright and obvious, even though it could be seen with the naked eye as a large and ghostly haze.

Before passing on to consideration of that very special object, Halley's comet, a few words should be said about the remarkable group of individuals who have gone "comet-hunting." Some really brilliant objects, such as the Great Comet of 1882, burst into view so dramatically that they were first noticed by casual sky-gazers. But a good percentage of comets have been discovered by dedicated observers who look for nothing else. The rest are picked up accidentally as hazy spots on photographs of the stars taken for some other purpose.

Serious comet-hunters tend to be amateurs, who spend countless hours scanning the night sky with big binoculars or telescopes. The late Leslie Peltier, of Ohio, discovered 12 between 1925 and 1954. Eric Alcock, observing from England, has found five between 1959 and 1983, his latest discovery the celebrated Earth-grazer already referred to. Bill Bradfield (Adelaide, Australia) has discovered 12 since 1971. In addition to the thrill of making such a discovery, there is also the satisfaction of having the comet named after you!

The comet bearing Halley's name is unusual, in that it was not discovered by him at all, although he did observe it. Halley's involvement was more "revolutionary," as we shall see.

480 Birth of St. Benedict, founder of Benedictine Order

500 High point of Aztec culture

(?)537 Death of King Arthur of Britain

530

3
The Story of Halley's Comet

Edmund Halley, who suggested that at least one known comet was a regular visitor to the Sun. (Drawing by Nigel Code.)

Edmund Halley was born in 1656, in a hamlet near London. As a student at Oxford University, he became interested in the motions of the planets around the Sun, which had been proved by Kepler, at

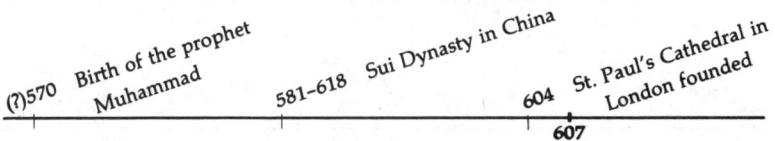

(?)570 Birth of the prophet Muhammad

581–618 Sui Dynasty in China

604 St. Paul's Cathedral in London founded

607

[1]

PHILOSOPHIÆ
NATURALIS
Principia
MATHEMATICA

Definitiones.

Def. I.

Quantitas Materiæ est mensura ejusdem orta ex illius Densitate & Magnitudine conjunctim.

AEr duplo densior in duplo spatio quadruplus est. Idem intellige de Nive et Pulveribus per compressionem vel liquefactionem condensatis. Et par est ratio corporum omnium, quæ per causas quascunq; diversimode condensantur. Medii interea, si quod fuerit, interstitia partium libere pervadentis, hic nullam rationem habeo. Hanc autem quantitatem sub nomine corporis vel Massæ in sequentibus passim intelligo. Innotescit ea per corporis cujusq; pondus. Nam ponderi proportionalem esse reperi per experimenta pendulorum accuratissime instituta , uti posthac docebitur.

B   Def.

Title page of Newton's Principia. *Halley saw it through the press, and paid for its production.*

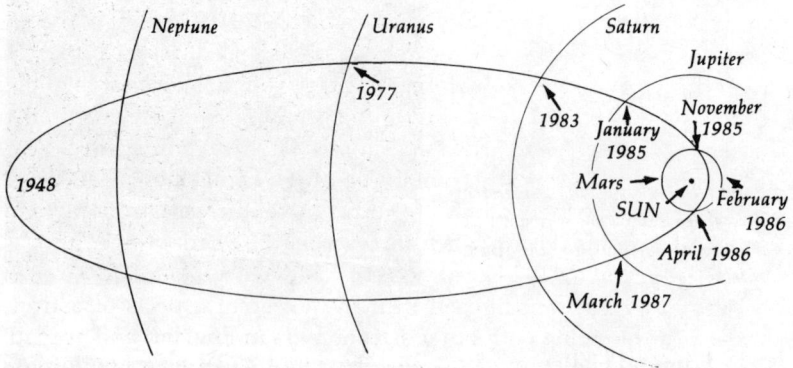

The immense extent of the orbit of Halley's comet can be appreciated from the fact that Earth's orbit is too small to be well represented on this scale.

618–906 T'ang Dynasty in China

651 Appearance of the Koran

673 Birth of the Venerable Bede, first English historian

684

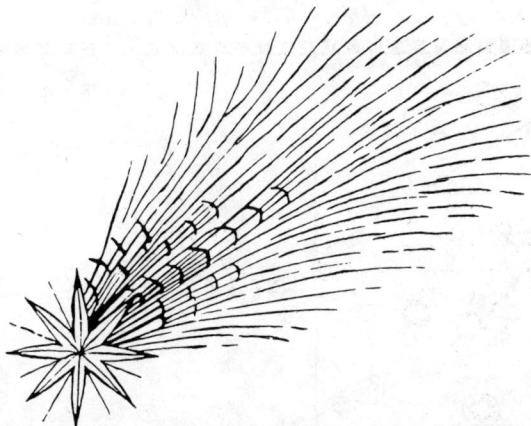

Halley's comet, as represented in the Nuremberg Chronicle at its return in the year 684.

the turn of the century, to be ellipses rather than circles. Since the theory of gravity had not yet been published, the *reason* for this elliptical motion was a mystery; and at the age of 19 Halley wrote an original paper on the velocities of the planets at different points in their orbits. With a wealthy father to support him, he was able in the following year (1676) to set sail for the island of Saint Helena, in the ocean wastes between Africa and South America, and spent a year cataloging the stars visible from the southern hemisphere. This work broke completely new ground for other adventurous astronomers to follow. Upon his return, when he was just 22 years of age, he was elected a Fellow of the Royal Society.

Halley interested himself in many different aspects of astronomy, corresponding with other astronomers in England and abroad, and visiting them, too—it is interesting to note that he made some important observations of the brilliant comet of 1680 while in Paris. But his attention was not exclusively on the stars. For example, in

(?)700 Composition of *Beowulf* (Nordic epic) begins

741 Death of Charles Martel

746 Outbreak of plague in Europe

760

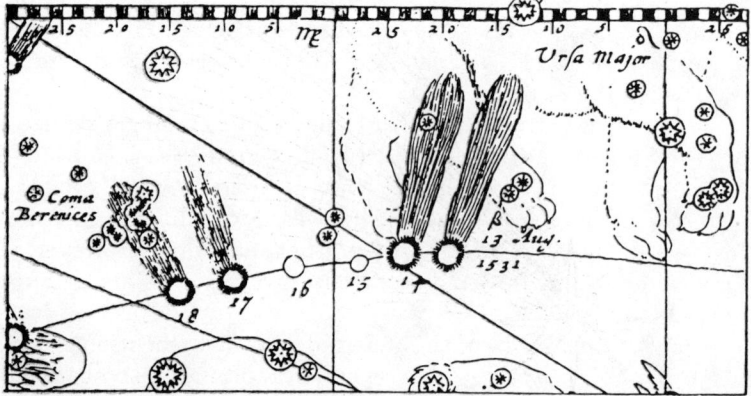

Halley's comet at its return in 1531. Some stars in Ursa Major (the Great Bear or Dipper) are seen at the right. These observations were used by Halley in his examination of the comet's orbit.

1694 he attempted to chart the variations of Earth's magnetic field by taking measurements on a journey to the South Atlantic; the first voyage ended in a mutiny, but on the second, in 1699, he was brought to a halt only when the ship encountered icebergs in the Frozen South. Other investigations were still less celestial; a biographer has commented:

Halley seemed equally at home writing to Leeuwenhoek in connexion with the latter's microscope observations on blood, the slime of eels, or cotton plants, and to Dr. Wallis at Oxford on such diverse topics as a "dropsicall maid", a child with six fingers and six toes, the kind of earth found at Hogsdon near London, and a host of astronomical matters.*

* Edmond Halley, 1656–1742. Memoir of the British Astronomical Association, Vol. 37, No. 3 (London: 1956).

762 Building of Baghdad commences

771 Origin of Arabic numeration

800 Charlemagne proclaimed Roman emperor

In 1703 he was elected to the Chair of Astronomy at Oxford, and in 1720 became the second Astronomer Royal. His last astronomical observations were made in 1739, three years before his death at the age of 85.

It is probably true to say that Halley's greatest single service to the world was to persuade his older contemporary, Isaac Newton, to publish his investigations into a law of gravitation, which appeared in 1687 as the *Principia*. In fact, he helped pay for the printing, since the Royal Society was practically bankrupt at the time! However, for the purpose of this book, his cometary investigations are of prime importance.

Halley was not so much interested in comets themselves as in their orbits. Unlike planets, whose elliptical orbits are so nearly circles that at first sight they could pass as such, comets usually have very elongated paths. Therefore, they afforded a good test of the gravitational theory, which needed to be seen to act at great distances from the Sun as well as in the nearby planet-inhabited region. Working on the assumption that a comet seen in some particular year could possibly be the return of the same one seen at a different epoch, moving in the same elongated orbit, Halley investigated the motions of 24 comets observed between the years 1337 and 1698, to see if the motions of any were compatible. When he published this work, in 1705, he commented:

Now, many things lead me to believe that the comet of the year 1531, observed by Apian, is the same as that which, in the year 1607, was described by Kepler and Longomontanus, and which I saw and observed myself, at its return, in 1682. . . . The identity of these comets is confirmed by the fact that in the summer of the year 1456 a comet was seen . . . and although its path was not observed astronomically, yet, from its period and path, I infer that it was the same comet as that of the years 1531, 1607, and 1682. I may, therefore, with considerable confidence, predict its return in the year 1758.

There was more to this hypothesis than simply noting the ap-

828 Egbert becomes first king of England

835 Appearance of first known printed book, in China

871 Reign in England of Alfred the Great begins

Plate XV

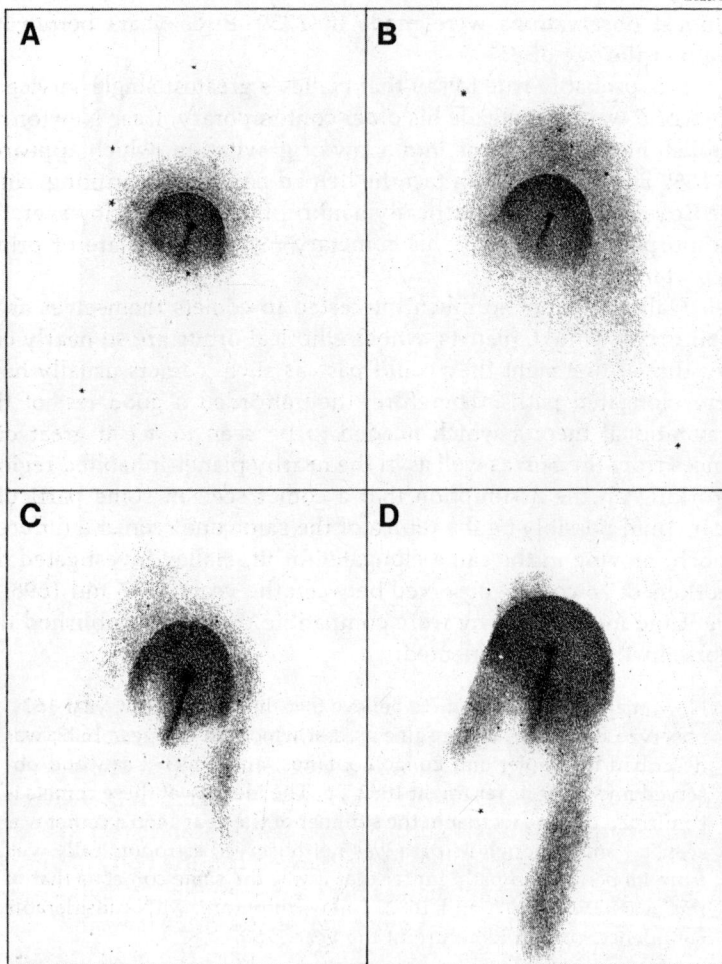

A

B

C

D

J F W Herschel del.

Hallmandel & Walton, Lithographers.

London: Smith, Elder & C° 65 Cornhill.

Telescopic drawings of Halley's comet at its 1835 return, made by the famous astronomer John Herschel on January 25 to 28. Only the coma and central condensation (with a bright "spine") are seen here. Each view covers an area of sky slightly smaller than that of the Moon.

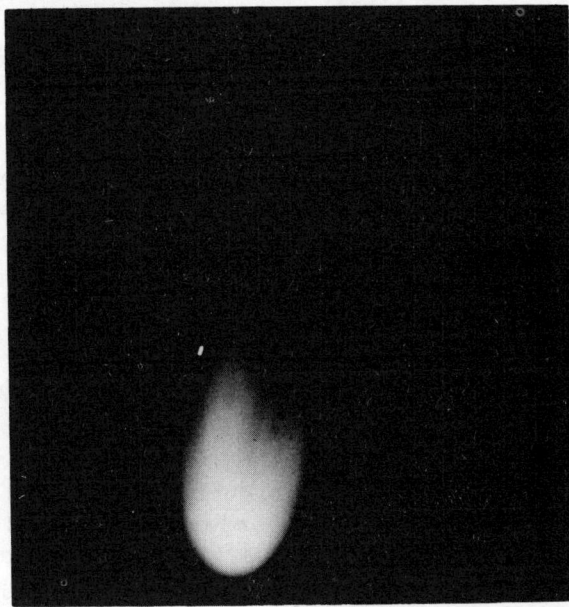

Halley's comet, photographed on May 10, 1910. The stars show as short lines, since the telescope followed the comet's motion during the 7-minute exposure. (Royal Astronomical Society, London.)

proximately equal periods of time between these cometary appearances. The work of turning the scattered, often vague, observations into an orbit round the Sun was a vast labor, which Halley had to carry out single-handed, and in longhand—remember, there were no calculators in those days! Furthermore, a significant effect known as "planetary perturbation" had to be allowed for. If a body of very little mass, such as a comet, passes near a planet, the gravitational pull of the latter can divert the comet out of the orbit it would pursue were it moving under the Sun's influence alone. The orbit of Halley's comet takes it near those of the giant planets Jupiter and Saturn,

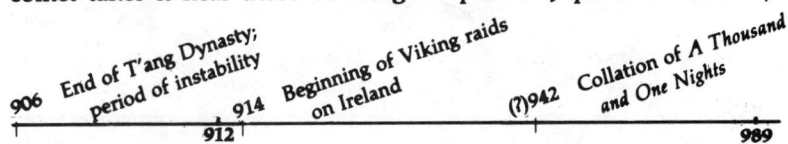

906 End of T'ang Dynasty; period of instability

912

914 Beginning of Viking raids on Ireland

(?)942 Collation of A Thousand and One Nights

989

which exert a considerable pull, so that important perturbations will occur. Halley had to develop his own methods of allowing for these perturbations, and discovered that the comet took more than a year longer to complete its approximately 76-year orbit between the returns of 1531 and 1607 and those of 1607 and 1682.

In making his bold prediction of a further return, Halley tried to allow for the pull of Jupiter, which would hold the comet back on its way to perihelion—without this effect, it should have appeared in the winter of 1757–58, but Halley suggested that a more likely time was the winter of 1758–59. As the time drew near for the posthumous return, the French astronomer Clairaut, after six months' intensive labor, calculated that the perihelion passage would occur as late as April 1759. The fact that the true date proved to be March 12, 1759, is an amazing one, considering that the comet had been invisible for three quarters of a century! Therefore, although Halley's wish that "If it should return, according to our predictions, about the year 1758, impartial posterity will not refuse to acknowledge that this was first discovered by an Englishman" has rightly been celebrated in the comet's name, it is regrettable that the wonderful computation of Clairaut and his assistants Lalande and Lepaute has tended to be overlooked.

Halley's comet, at this epoch-making return, was first sighted on Christmas Day, 1758, by a farmer and amateur astronomer named Palitzsch, observing near Dresden. Conditions during that passage were rather similar to those that will occur at the 1986 return, for the comet will pass perihelion only a month earlier than it did in 1759. It was followed by many observers, including Charles Messier, one of the greatest of all comet-hunters, until it vanished into the evening twilight in the middle of February 1759. After perihelion it reappeared in the east, rising before dawn, and in April and May was a fine object to observers in the southern hemisphere, as it is expected to be during March and April on the coming occasion.

It may be of interest to briefly chronicle all the recorded appear-

(?)1001 A shipload of Greenlanders reaches the American continent

1016 Canute becomes King of Denmark, Norway, and England

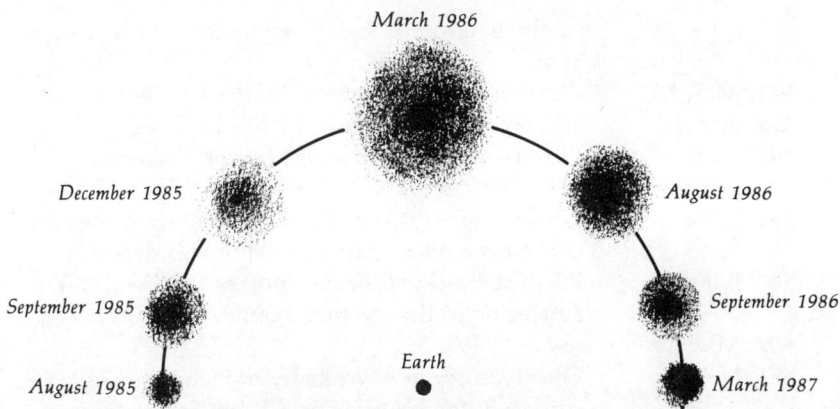

How the size of its coma is expected to change when Halley's comet passes through perihelion.

ances of Halley's comet, with a longer description of its most recent return, in 1909–11. Each entry is preceded by the date of perihelion passage.

B.C. 239, May 15   Chronicled by the Chinese.

B.C. 162, May 20   Some authorities dispute that these records refer to Halley's comet at all.

B.C. 86, Aug. 15   Chronicled by the Chinese.

B.C. 11, Oct. 9   Observed for nine weeks by Chinese astronomers. Seen from Rome before the death of Agrippa.

A.D. 66, Jan. 26   Observed for seven weeks by the Chinese, and possibly alluded to by the Jewish historian Josephus, describing the destruction of Jerusalem by Titus: "Thus there was a star resembling a sword which stood over the city, and a comet that continued a whole year"; though Halley's comet cer-

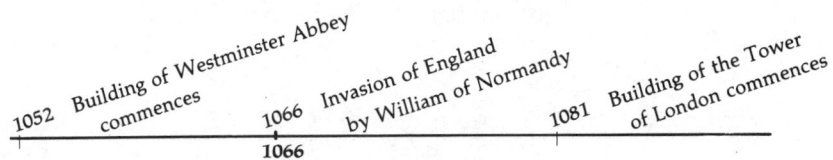

1052 Building of Westminster Abbey commences

1066 Invasion of England by William of Normandy

1081 Building of the Tower of London commences

1066

tainly could not have been seen for a whole year!

141, Mar. 25	Observed for four weeks by the Chinese.
218, Apr. 6	Observed for six weeks by the Chinese.
295, Apr. 7	Observed for seven weeks by the Chinese.
374, Feb. 13	——
451, Jul. 4	Observed for 13 weeks by the Chinese. Appeared at the time when Attila the Hun was defeated.
530, Nov. 15	Recorded as "very large and fearful," with a tail extending to the overhead point, or zenith.
607, Mar. 20	——
684, Nov. 6	Observed for five weeks by the Chinese, and recorded in the Nuremberg Chronicle.
760, Jun. 11	Observed for eight weeks by the Chinese.
837, Feb. 25	Observed for five weeks by the Chinese. This is claimed to have been the comet's brightest recorded appearance.
912, Jul. 19	Recorded by Japanese astronomers.
989, Sep. 2	Observed for five weeks by the Chinese, and also recorded in Anglo-Saxon annals.
1066, Mar. 27	Observed by the Chinese for 67 days. It coincided with the Norman invasion of England, and its appearance is recorded in the Bayeux Tapestry. The first observation was not made until April 2, a week after perihelion passage.
1145, Apr. 19	——
1222, Sep. 10	English observations noted the comet as appearing like a bright star with a tail.
1301, Oct. 23	Visible for six weeks, and observed in China and in Europe.
1378, Nov. 9	Observed by the Chinese for six weeks.
1456, Jun. 8	According to Chinese observations, the tail resembled a peacock's and appeared up to 60° long (remember that your outstretched fingers, at

1096   The First Crusade

1155   Birth of Genghis Khan,

1145

arm's length, subtend about 20°). Observed for four weeks, in Europe as well as in China.

1531, Aug. 26   Visible for five weeks.

1607, Oct. 27   Visible for nine weeks. This was the last of the comet's returns before the invention of the telescope, and it was first detected by Kepler.

1682, Sep. 15   The tail had a maximum recorded length of 30° at this return, and in England it "appeared in great splendour." This was the apparition observed by Halley himself, who made telescopic observations and saw bright rays or jets in the head of the comet. It was followed for five weeks, and first sighted by Flamsteed, the first Astronomer Royal.

1759, Mar. 13   The first predicted return. The best display was seen after perihelion; on May 5 the tail was reported to have been 47° long. It was followed for five months.

1835, Nov. 16   The comet was observed over a period of 41 weeks, from August 5, 1835, until the last sighting on May 5, 1836. The greatest extent of tail, about 20°, was attained on October 15.

1910, April 20

This apparition was observed very intensively, since photography had been invented since the previous return, and more powerful telescopes were available. The comet was first detected in August 1909, when 500 million kilometers from the Sun, or midway between the orbits of Mars and Jupiter, and was followed until June 1911. By that time, it had receded into the chill depths of space beyond Jupiter.

1170 Murder of Thomas Becket in Canterbury Cathedral

1215 Magna Carta signed

1226 Death of St. Francis of Assisi

1222

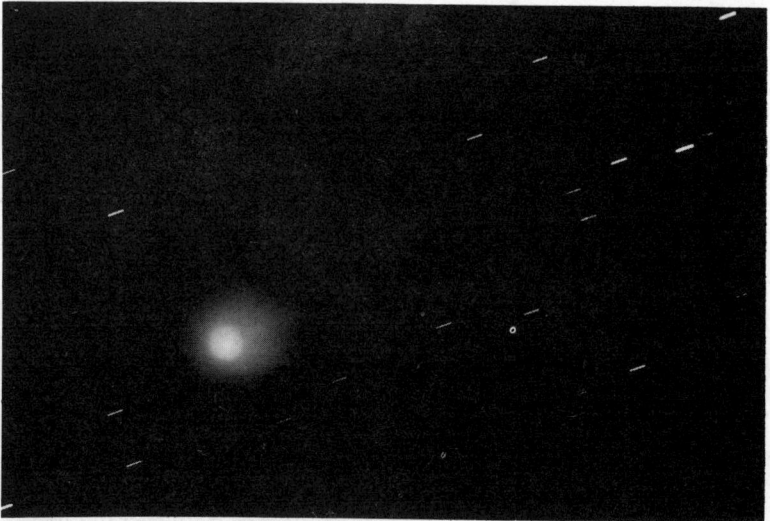

Halley's comet, photographed on June 2, 1910. The exposure was 30 minutes. Note the fine filaments in the tail, about 1° of which is shown here. (Royal Astronomical Society, London.)

We should, first of all, clear up a confusion over what happened in 1910: for Halley's was not the only comet to appear in that year. In the middle of January, a brilliant comet burst into view. Having approached the Sun from the side opposite to that on which Earth was located, it escaped observation until practically at perihelion, which occurred on January 17. At this point it was visible with the naked eye, although only 4° away from the Sun's disc, as a brilliant spot with a short tail against the blue daylight sky! At sunset, the tail streaming up from the horizon was seen by many people. However, the comet faded more rapidly than expected, and did not become the truly "great" comet of its sudden promise. Nevertheless, in the minds of those now elderly people recalling the events of 1910, the

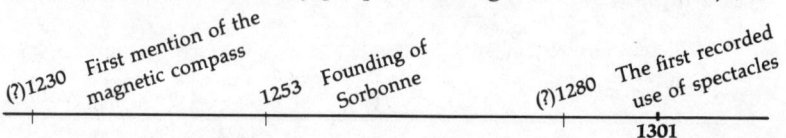

(?)1230   First mention of the
          magnetic compass
                              1253   Founding of
                                     Sorbonne
                                                   (?)1280   The first recorded
                                                             use of spectacles
                                                             1301

Daylight Comet and Halley's comet may have become confused.

Halley's comet was not particularly conspicuous before perihelion, but, following its normal pattern of behavior, it grew a much longer tail afterward. To the casual eye, its maximum extent was about 30°, but skilled observers benefiting from the clear, dark air of mountaintops were able to detect it for 100° or more. If we turn these angles into real lengths in space, the maximum tail development was of the order of 100 million kilometers, which is a large proportion of Earth's distance from the Sun! Even the comet's coma attained a diameter of some 150,000 kilometers, which is about 12 times the diameter of Earth.

This vast size seems utterly unbelievable when it is appreciated that the solid nucleus of Halley's comet, from which the coma-making and tail-making material is derived, measures only about five kilometers across. This reflects, more clearly than any numbers, the fantastic tenuity of the visible part of a comet. It is visible at all only because our eyes can take in, at a glance, the contents of millions of cubic kilometers of space. To match the density of a comet's tail, one teacupful of air would have to be expanded to fill the volume of the whole Earth.

In physical terms, therefore, Halley's comet is as nearly nothing as anything could be. This was well demonstrated on the two interesting days of May 18 and 19, 1910. On the first day, the head of the comet passed between Earth and the Sun, but despite the most careful search no dark outline was observed on the solar disc. On the following day, according to calculations, Earth passed through the outer regions of the tail itself. Unfortunately, the Moon was near full, which made it difficult to detect any sky glows that might have been caused by the passage; but our planet certainly suffered no noticeable ill-effects. It was hardly supposed that it would. In 1861, Earth had passed through the tail of another comet, an event betrayed only by a curious phosphorescence of the night sky.

The use of photography, and the work of some very careful observers, allowed much detail in the coma and tail to be made out, and

Beginning of the Hundred Years' War 1337

1348 Boccaccio commences work on the Decameron

it was realized that bright clouds of material were pouring out of the nucleus at a great rate. Noticeable changes could be seen over a period of an hour or two, and it was calculated that one particular fragment of tail had been ejected at a speed of some 50 kilometers per second! However, none of this violent activity would be noticeable to a casual glance.

By the time the comet was lost from view, using the largest telescopes then in existence, it appeared as a tiny, starlike speck on a photographic plate, only about one millionth of the brightness it had when it was most prominent. By 1982, when it was recovered again on its way to the forthcoming perihelion, it was about 1,600 times fainter than when last detected in 1911! Such is the progress astronomical detecting instruments have made during this most recent beat of the Halley clock.

1368–1644 Ming Dynasty in China

1382 First translation of the Bible into English, by Wycliffe

1378

4

Observing Halley's Comet

A comet has two sides to its nature. There is the predictable one, to begin with. Its motion through space is subject to the laws of gravity, which determine the physical attraction between any two bodies, dependent upon their masses and the distance separating them. Knowing these values for the Sun, the comet, and any disturbing planets, it is possible to calculate the comet's orbit through space, and hence its apparent path in front of the stars, as seen from Earth's surface. The tables and charts in the next chapter are not expected to be in error as far as this information is concerned.

What cannot be predicted so accurately, however, is what the comet will look like. Its brightening and fading as it passes perihelion, and the development of its coma and tail, vary from one apparition to another. This may partly be due to alterations in solar activity; the Sun passes through an 11-year cycle of change, and since comets are heated and made active by the Sun's radiation, solar activity could have an effect upon their "performance." It must also be remembered that the geometry of Earth and Halley's comet is never repeated exactly, and so our view of its course in toward the Sun and away back into space can be good or poor, depending upon the date of perihelion. It is interesting to note that the current perihelion date (February 9) is only two weeks away from the perihelion passage of A.D. 837, February 25, which is claimed to have been the comet's brightest appearance on record. However, its brightness in 1986, based on its appearance at the last three returns, is not expected to be anything like that of 837, when it appeared almost as

1400 The Canterbury Tales left unfinished at Chaucer's death   1415 Battle of Agincourt, where Henry V defeats the French   1431 Joan of Arc burned at the stake

The approximate position of Halley's comet in the sky, as seen by an observer facing south at the following times:

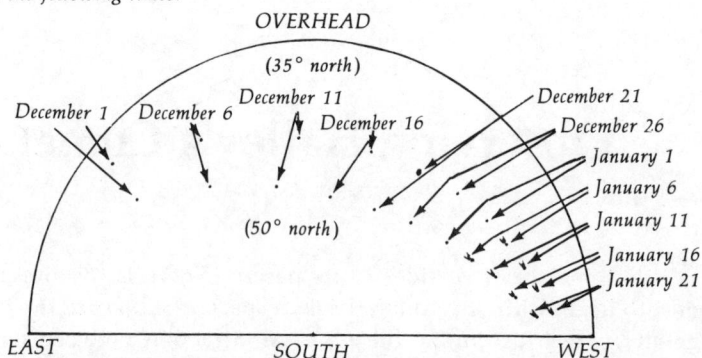

(a) At the commencement of darkness (end of astronomical twilight) on evenings between December 1, 1985, and January 21, 1986.

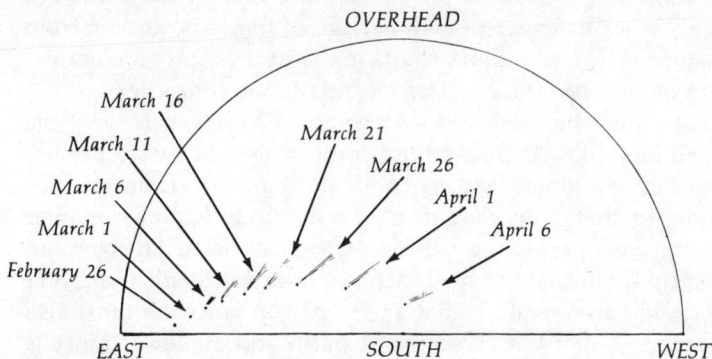

(b) At the end of darkness (beginning of astronomical twilight) on mornings between February 26 and April 6, 1986. (Latitude 35° north only.)

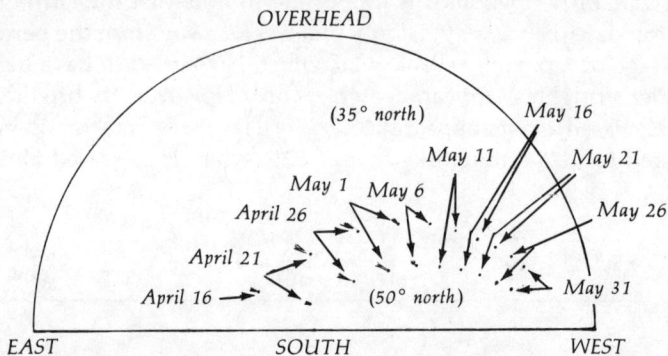

(c) At 2200 hours local time on evenings between April 16 and May 31, 1986.

bright as the planet Venus. Could the comet be growing intrinsically fainter as more and more material is blasted off into space with each perihelion passage?

The accompanying diagrams reveal the approximate position that Halley's comet will have in the night sky, as seen from our two standard latitudes of 35° and 50° north, during the period from the beginning of December 1985 until the end of May 1986. More detailed notes are given in the following chapter, but the following summary may be helpful to give an initial picture.

Observer at 50° North

During the month of December, when the comet should brighten from easy binocular to possible naked-eye vision, it will be fairly well placed for observation in the earlier part of the night. During the first half of January it will brighten further and may develop a short tail, but will disappear into the evening twilight, in the west, round about the middle of the month. It will probably not be seen again until the second half of April, when it will be past its best and sinking below naked-eye visibility again, located in the pre-midnight sky.

Observer at 35° north

The December view will be rather similar to that described in the preceding, but the comet will be more favorably placed in January, and may be followed for a few days longer. The most important benefit of being at this lower-latitude site will be noticed after perihelion, particularly during March, when the comet will rise in the early hours. Although it will be low, the tail should be easily seen. It may be lost for a week or two in the first half of April, but thereafter it will be seen rather better than from a latitude of 50°.

Viewers located in the southern hemisphere will have a less sat-

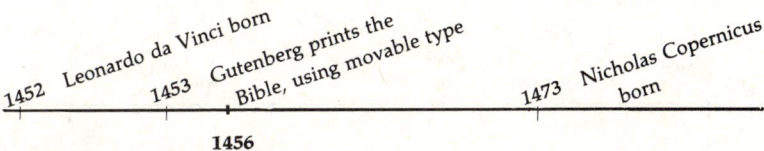

1452 Leonardo da Vinci born   1453 Gutenberg prints the Bible, using movable type   1473 Nicholas Copernicus born

1456

A pair of 11 × 80 binoculars made by Celestron. Binoculars of this power will enable Halley's Comet to be followed for several months.

isfactory sight of the pre-Christmas binocular phase, but their situation will be ideal for seeing the "tailed" phase in March and April 1986. Anyone wishing to enjoy Halley's comet to the full is strongly advised to travel south of the equator!

Instruments for Observing Halley's Comet

It is worth remembering that 25 of the 29 returns recorded in historical annals were observed with nothing but the naked eye. So, don't despair if you have no optical equipment at all, unless you happen to live in the higher of our two northern latitudes, where you may have difficulty seeing the comet at all, either before or after perihelion. This will apply particularly if you live in an urban area, where the comet may be a "nonevent" if you base your expectations

1475 Michaelangelo born   1492 Columbus lands on Watling Island   1517 Luther's 95 Theses

1531

on the fine photographs of the 1910 return, or on the reported appearance of some of the brighter comets of more recent times.

It makes sense, therefore, to have at least a pair of binoculars at hand. Even if the comet can be made out without any optical aid at all, a small instrument will vastly improve the view. Any binoculars will be of some service, but the following notes may be helpful when making a selection.

Magnification

Every pair of binoculars has a specification marked on it: two numbers separated by an ✕. The first of these numbers is the magnification, or how many times higher or wider an object appears when viewed through the binoculars. The most common magnification is 8✕, but powers of from 6✕ to 12✕ are widely available. Don't assume that the highest magnification is automatically the best. Very faint extensions of tail and coma tend to disappear if too high a magnification is used, while the amount of sky visible at one time (the field of view) becomes smaller as the magnification increases. In general, and certainly during the comet's brighter phase, you will want the view to include as much of the tail as possible. A magnification of 8✕ or 10✕ will probably prove the best choice.

Aperture

The number after the ✕ is the aperture or diameter of the large lenses at the front of the binoculars, in millimeters. The bigger these lenses are, the brighter the image will be. This is because a large lens collects and focuses more light energy than a small one. A popular aperture is 30 mm, but for astronomical work, where light is at a premium, you will appreciate the luxury of 50-mm lenses, and 70-mm binoculars can also be obtained quite easily. A pair of 10✕50 or 10✕70 binoculars will be excellent, but even the common 8✕30 type

1538 World map published by Mercator

1564 William Shakespeare born. Galileo born

will give a far better view than the unaided eye, as you can prove to yourself by viewing the night sky with them on any convenient occasion.

Quality

Cheap binoculars can be a headache—literally! If the two optical sets that separately serve the eyes are not in good mutual alignment, or "collimation," the two images will not merge in the brain without a good deal of strain, and your eyes will rapidly become sore and tired. Cheap binoculars can easily come out of adjustment, and, indeed, may never have been in good adjustment to begin with. Furthermore, they will not give as sharp an image as will a good pair: certainly not right across the field of view. It is a common experience to find stars at the center of the view to be reasonably "starlike" (in other words, looking like little points of light, and *not* with long streaks extending on all sides!), while those near the edge of the field of view are extended fuzzy patches. This is an irritating characteristic of cheap instruments. If a star, even when it is near the center of the view, cannot be focused down to a sharp point without flare or colored fringes, then consign the binoculars to the trash can.

Focusing

It may seem surprising to devote a paragraph to this elementary necessity, but a surprising number of binocular users are not aware of how to get the best out of their instruments in this way. Eyepieces have to be adjusted from time to time, not only to compensate for the vision of the particular user but also because objects at different distances have their own setting of best focus. Some binoculars have separately focusing eyepieces, which have to be twisted separately to get the image sharp—these are often abbreviated "IF" for "independent focusing." These are ideal for astronomy, since they are partic-

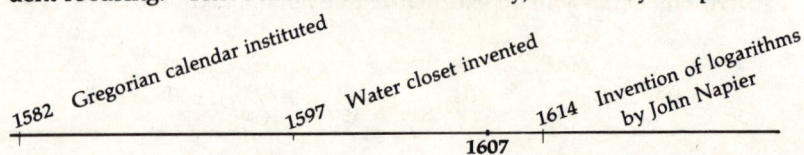

1582 Gregorian calendar instituted

1597 Water closet invented

1614 Invention of logarithms by John Napier

1607

ularly stable, and there is no need to refocus on objects at different distances. But the vast majority of models are "CF" or "center focusing." These have a central wheel, to adjust both eyepieces simultaneously, which is all right provided that the eyepieces are in the correct relative adjustment to start with. If you take a look at a CF model, you will find that one eyepiece (usually the right-hand one) has its own separate adjustment, while the left-hand one cannot be twisted at all. So the procedure is to focus the *left* eye by using the center wheel, and to focus the *right* eye after this is completed by twisting the right-hand eyepiece. This will bring the two eyepieces to their correct relative adjustment, and any future focusing can be achieved by using the center wheel alone.

Using an Astronomical Telescope

You may have access to something larger and more powerful than binoculars—in short, a proper astronomical telescope. There is not room to discuss telescopes here. If you have such an instrument, it is likely that you already have some experience of observing, and have purchased an observing guide (if you haven't, the writer's *Amateur Astronomer's Handbook*, published by Harper & Row, contains plenty of information). But it may be worth mentioning that a pair of binoculars will, almost certainly, give a better view of the whole comet when it is in its March–April "tailed" stage, simply because they offer a much lower magnification than is possible with an astronomical telescope. A low magnification means a wide field of view, so that more of the comet can be seen at one time.

Preparing to Observe

Some tips about observing may be in order if you have never done anything of the sort before. First, wrap up warmly—a March dawn can be bitter indeed. Observer comfort is critical: if you are

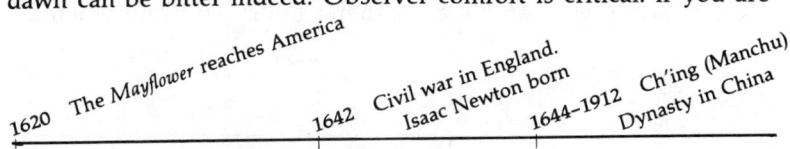

The Mayflower reaches America

1620

1642 Civil war in England.
Isaac Newton born

1644–1912 Ch'ing (Manchu)
Dynasty in China

cold, hungry, and tired, the results will not be very satisfactory! If you are reluctant about getting up in the small hours of the morning, it is good psychology to have your observing clothes sorted out before going to bed, and to plan on having a pot of coffee handy to give yourself the necessary encouragement. One's attitude toward astronomical observing can change radically in the course of a night!

This may sound like common sense, and so it is. Less obvious is that you cannot hope to see very much of the comet, regardless of the sky conditions, without allowing your eyes to acquire that particularly sensitive state known as *dark adaptation*. Even the brightest stars may be hard to see, if you come straight out of a dazzling room. But remain in the darkness for a few minutes, and a minor miracle happens: a sensitizing secretion known as *visual purple* coats the retina (the screen at the back of the eye, where the image is formed) and renders it hundreds of times as sensitive as it is in normal bright conditions. If the eye were left in this supersensitive state all the time, we would become completely blinded in daylight. So do not be disappointed if you can see little or nothing of the comet during the first exciting minutes as you try to locate it. The first stage of dark adaptation takes some five minutes, but night vision continues to improve significantly for perhaps half an hour. To illuminate the chart or notebook, use a dim red flashlight—and stand in the shadow of any lighted windows or street lamps, for they will certainly not improve your sensitivity to faint illumination.

Generally speaking, it is not worth the attempt to observe any celestial object when its altitude above the horizon is less than 10° or so. At these very low altitudes, the effective thickness of air through which you are viewing a celestial object is considerably increased. Air absorbs light; therefore the object appears much fainter. Furthermore, if you are in or near a town, you will find that low altitudes are very hazy, and glow with diffused street lighting. In the notes, however, we give details of the comet's rising and setting times on selected dates, not implying that it magically sweeps into or out of view at such a moment, but as a general guide to the time when it is

1660 Restoration of English monarchy

1682 Founding of Pennsylvania

1682

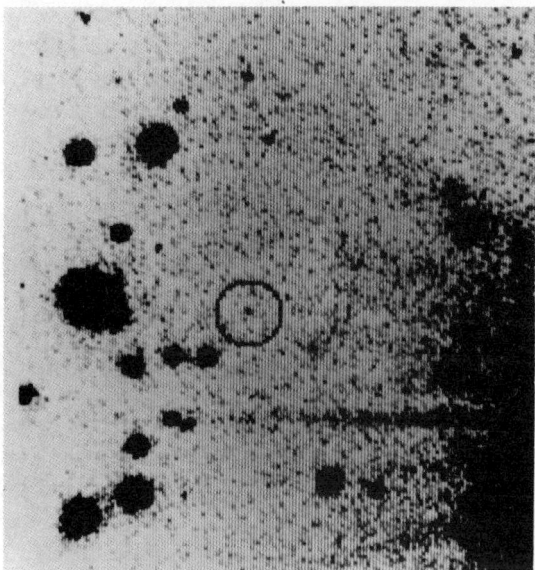

The photograph on which Halley's comet was recovered, taken on October 16, 1982. This is a reverse print, on which stars appear black. It was then about a hundred million times fainter than the dimmest naked-eye star. (Palomar Observatory Photograph, Caltech. Taken by Doctors David Jewitt and G. Edward Danielson.)

above the horizon. There is also the possibility that, during the generally difficult post-perihelion stage of March and early April, you may be able to see part of the tail extending up above the horizon haze, even if the head itself is too low and obscured to be made out.

Things to Look For

What you do once you have found the comet, is a matter of choice. You may decide to keep a record of the brightness of the

1705 First workable steam engine devised by Newcomen

1732 Birth of George Washington

1759

central condensation and coma combined, by comparing it with nearby stars. To do this, you must study the appearance of the head, and then put the binoculars sufficiently out of focus so that the nearby stars appear to have enlarged to a similar size; then make your choice. Subsequently, the stars you have used to make your brightness estimate can be identified from the maps in this book, or from a star atlas, and their brightness, or *magnitude*, determined from a catalog. Some useful publications are mentioned in Chapter 5 (see page 45), and you will find helpful hints on how to make magnitude estimates in *Astronomy with Binoculars*.

Another interesting project will be to document the changing appearance of the tail, whether seen with the naked eye or with binoculars, drawing its shape and extent (marking in nearby identifiable stars, for reference) and looking for any fine structure visible within it. If you are using a more powerful instrument, you can study any bright jets and envelopes in the coma, which could show change in the course of a single session. You can, if you wish, plot as accurately as possible the path of the central condensation in front of the stars, and keep an eye out for any "transits" of the comet in front of stars. Typically, stars can be seen shining right through comets, which is further testimony to their "nothingness"—consider how readily even bonfire smoke can obscure the Sun, and yet a star can penetrate something a hundred thousand kilometers thick! You will, probably, want to extend your period of detection for as long as possible, into the months of June and even July, if you have a favorable site and a large instrument.

Photographing Halley's Comet

The forthcoming apparition of Halley's comet is going to be observed as it has never been observed before. Countless photographs will be taken, to add to the pile that started with the discovery

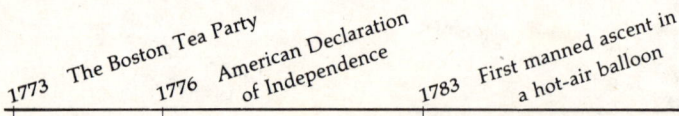

1773 The Boston Tea Party

1776 American Declaration of Independence

1783 First manned ascent in a hot-air balloon

picture, secured on October 16, 1982, when it appeared *one hundred million* times fainter than the faintest stars visible with the naked eye. However, many of these photographs will be obtained using special large instruments, and will give no idea of the comet's visual appearance, especially when observed with the naked eye. Such photographs, showing the comet as a large section of the population will remember it, are the province of the owner of an ordinary 35-mm camera. Even if you have never photographed anything of the kind before, it will be well worth trying.

Astronomical photography differs from ordinary photography in several ways. It may seem surprising that a star or a comet can be photographed at all without special equipment. But it can be done, and has been! The following points need to be kept in mind:

1. Most astronomical objects are relatively faint, and exposures need to be much longer than those used in routine photography. Highly sensitive film is almost always required.

2. Many astronomical objects (particularly stars) are of a great range of brightness. The longer the exposure, and the more sensitive the film, the greater the number of stars recorded. Therefore, there is no single "correct" exposure for a particular piece of sky: a short exposure records the brighter objects, and a long exposure brings in fainter objects as well.

3. Earth is spinning. If you fix your camera to Earth by standing it on a tripod, or laying it on the ground, it will partake in this rotation, and the stars (or Halley's comet) will trail across the film. Therefore, with a so-called "fixed" camera, there is a limit to the usable exposure before the celestial objects show trailing. To eliminate this effect, the camera must be attached to a so-called *equatorial mounting*, which takes us into the province of the amateur astronomer. We shall assume a fixed camera.

Using a 35-mm camera with a standard lens set to its maximum opening, and loaded with the most sensitive film available, all the naked-eye stars can be recorded with an exposure of about one sec-

1789 The French Revolution

1804 Napoleon proclaimed emperor of France

1815 Battle of Waterloo

ond. An exposure of about 20 seconds is near the useful limit; anything longer will begin to show the effects of trailing, which is indeed a comment on the rate at which our planet spins! With a 20-second exposure, you can expect to record stars at approximately the limit of a pair of 8×30 binoculars, assuming that the sky is dark and transparent, with no moonlight or obtrusive artificial glare. Therefore, photographers have the potential to record Halley's comet, without using any special equipment at all, from roughly the beginning of November 1985 to the middle of June 1986, provided it is well up above the horizon as seen from their site. The following hints may be useful:

Choice of Film

Color transparency film tends to give the most pleasing results, since the contrast of a slide is more brilliant than that of a print. Ektachrome 400 is recommended. For black-and-white work, Tri-X is a good choice.

Lens Adjustment

Set the lens to its widest aperture, whatever this happens to be. If in doubt about which way to twist the collar, look into the front of the lens and observe the expandable diaphragm. Make sure that the lens is focused on infinity!

Shutter Setting

Modern cameras rarely have the old-fashioned "Time" shutter setting, which meant that the lens could be left open indefinitely. There is usually only a "B" setting, which means that the shutter remains open while the release is pressed, but closes immediately when the pressure is removed. Therefore, obtain a cable release with a locking screw, so that the shutter may be held open.

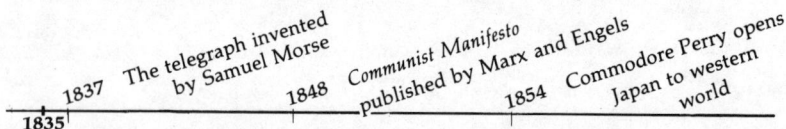

1835 | 1837 The telegraph invented by Samuel Morse | 1848 Communist Manifesto published by Marx and Engels | 1854 Commodore Perry opens Japan to western world

Bennett's comet, which made a fine display in 1970, photographed by the writer using a fixed 35-mm camera and Tri-X film, exposure 20 seconds

Aiming

Direct the camera, so that it is held rigidly, toward the part of the sky known to contain the comet—which need not necessarily be visible to the unaided eye. It is not always easy to see stars through a camera viewfinder, and you may have to judge the direction of aim. With ingenuity, a simple frame viewfinder can be constructed from wire, and painted white to show up better in the dark.

1861   American Civil War begins

1865   Abraham Lincoln assassinated

1879   The electric lamp invented by Edison

Camera Protection

Be sure that the camera lens is shaded from any artificial light. Even if you have an agreeably dark site, it is always a good idea to use a lens hood to obtain extra protection. If the weather is damp, check that no dew has formed on the lens; inspection with a flashlight (lens shutter closed!) will reveal any telltale haziness. Dew will not form on any surface that is warmer than the air, so that, if conditions are "dewy," it is a good idea to keep the camera in a warm room until needed.

Exposure Length

Decide upon the exposure. With a fixed camera, there is little point in giving less than the maximum "nontrailing" amount, but if these pictures are your first effort at astrophotography it will be instructive to experiment with a series—say, 1, 2, 4, 8, and 16 seconds—and to compare the resulting star and comet images.

Making the Exposure

Now you are ready to take your picture. Do not just snap the shutter open and shut. The chances are that the camera will vibrate, even if you are using a cable release, since most models have a focal plane shutter—a kind of blind that flies across in front of the film with a bang. The best technique is to cover the lens with a piece of black card, and *then* to open and lock the shutter. Now gently lift the card away from the lens, and replace it after the required time interval. Finally, close the camera shutter.

Lenses

If you have a choice of lenses for your camera, then by all means experiment. The standard lens for a 35-mm camera has a

1896   The Olympic Games revived

1897   Theodor Herzl founds first Zionist World Congress

1901   The first successful radio transmission across the Atlantic, by Marconi

focal length (the distance from the lens to the film) of about 50 mm. Such a lens will show an area of sky measuring about 40° × 27°, which means that it could quite easily include all the bright stars in the Big Dipper or Great Bear. If you substitute another lens with, say, a focal length of 90 mm, the area of sky shown will be reduced to 23° × 15°—the size of Orion, the Hunter. In other words, the same object will appear larger on the film. You will find that the stars and comet will trail more quickly, so that the exposure has to be reduced, but the larger image size may compensate for this.

If you happen to know someone with a properly mounted astronomical telescope, take your camera around to his yard, strap it to the instrument, and try some guided exposures. The results may inspire you to embark on a new and richly rewarding hobby, which will last long after the subject of this book has rounded the Sun and returned once more to its obscurity. After all, the stars and the planets, the Sun and the Moon, are always there; and a lifetime is not long enough to exhaust their fascination. Whether you photograph, or draw, or just look, you may come to thank Halley's comet for focusing your attention on the unsuspected universe that envelops the denizens of our little world. Good luck!

1903 Wilbur and Orville Wright make first successful airplane flight

1909 Peary reaches the North Pole

1910

5

Where to Find Halley's Comet

(DECEMBER 1, 1985—JULY 1, 1986)

Between the months of December 1985 and May 1986, Halley's comet will be near enough to Earth and the Sun to be visible from some parts of the world with the naked eye or with the slightest optical aid. This chapter gives month-by-month instructions for locating it during this time, and continues on into the period when it will be an exclusively telescopic object.

The latitude of the observer can have an important effect upon the visibility of the night sky, and therefore of the comet: this has already been referred to on page 29, where the two standard latitudes of 35° north and 50° north were introduced. This band of Earth's surface corresponds approximately to the location of most European and American observers, but with a little judgment the notes can also be used by people on either side of these latitudes. The general principle is that the further north you live, in Earth's northern hemisphere, the worse the view; the further south you live, the better the view.

Each month's notes give the following assistance:

1. General information, with a small-scale key chart, on the comet's whereabouts in the sky, and at what time of night to look for it.
2. A more detailed "close-up" view of its position, usually at five-day intervals, in relation to nearby stars and constellations. In most cases, the map shows all the stars that are readily visible with the unaided eye in good sky conditions; if the region is low in the sky, however, the fainter ones will not be seen without binoculars. The brighter the star, the larger is its disc on the map.

43

3. Some information about the comet's likely brightness and appearance.

4. A small diagram showing the movement of the comet, and Earth, along their respective orbits during the course of the month, and some information about the distance separating them.

5. A table giving the times of the comet's rising and setting, as well as when it is on the meridian (due south in the sky). Celestial objects are always at their greatest altitude above the horizon when on the meridian. Many of these times will occur in daylight or bright twilight, when the comet cannot be observed anyway; but it may be interesting and useful to know its position in the sky at any given time. The time when "astronomical twilight" begins (at dawn) or ends (at dusk) is also given, as a guide to when the sky should be completely dark. Note that when the comet is at its brightest, observations may be possible when the sky is still perceptibly twilit, and this will be important when the comet is in the region of the Sun's glare, around the time of perihelion. These times are all given using the 24-hour clock, and are in local time, which may differ by up to half an hour or so from the standard time in civil use. Boldface type in a table means that this particular phenomenon occurs in a dark sky; ordinary type means either full daylight or twilight. Remember to allow for summer time or daylight saving time, if in force.

6. Appended to this table, a column headed "Altitude." With the exception of April (see pages 64–66), this indicates the altitude of the comet above the horizon at the beginning or ending of astronomical twilight, as appropriate. This will give you some idea of your chances of actually sighting the comet, bearing in mind that even bright objects are hard to make out within 10° or 15° of the horizon. But it should not be forgotten that the comet's tail, which points away from the Sun, will be at a greater altitude than the head, and could possibly be visible even if the head is obscure.

7. During the brighter phase (December–April), some space has been included for you to write up your observing notes month by month, so that this book may become a permanent, personal record, perhaps to be handed down to the generation that will witness the next return of Halley's comet, in the year 2062!

If you wish to supplement this book with other references, consider the following:

Norton, A. P., *A Star Atlas and Reference Handbook*, 17th ed. Edinburgh, Scotland: Gall & Inglis, 1978. This most useful volume shows all the naked-eye stars visible in the whole sky, and includes useful tables and information.

Tirion, Wil, *Sky Atlas 2000.0.* Cambridge University Press, 1981. This atlas shows stars down to the faint binocular level, and is becoming a standard for the enthusiastic amateur.

Muirden, James, *Astronomy with Binoculars.* New York: Arco Publishing, 1984. This book describes the wide range of opportunities offered by binoculars for astronomical work—including comet observation!

Yeomans, Donald K., *The Comet Halley Handbook.* NASA, 1983. A most comprehensive publication, with daily tables of the comet's position up to March 23, 1987.

You may also find a planisphere useful, particularly if you live in a latitude not served by this book. The best-known series is published by George Philip & Son, London. A planisphere can be "dialed" to show the sky as it appears, from a given latitude, at any time on any night of the year.

D E C E M B E R

Dipper   Pole Star

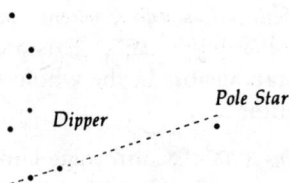

General Location

In Pisces at the beginning of the month, passing into Aquarius on the twenty-first. On the first it will not be setting until well after midnight, but by the end of the month it should be hunted as soon as the sky becomes dark.

To Find the Map Region

1. Locate the Pole Star. If you are not familiar with it, use the two "pointers" in the Dipper, as shown in the diagram. It is fixed in the north direction, and its altitude above the horizon is equal to your latitude on Earth's surface.
2. Lie on your back with your head to the north, and find Cassiopeia, which looks like a W near the overhead point, or zenith.
3. Referring to the key chart, strike a line from the Pole Star through

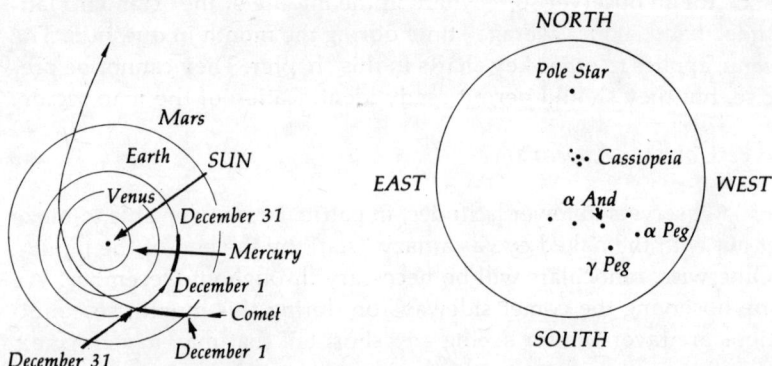

1 9 8 5

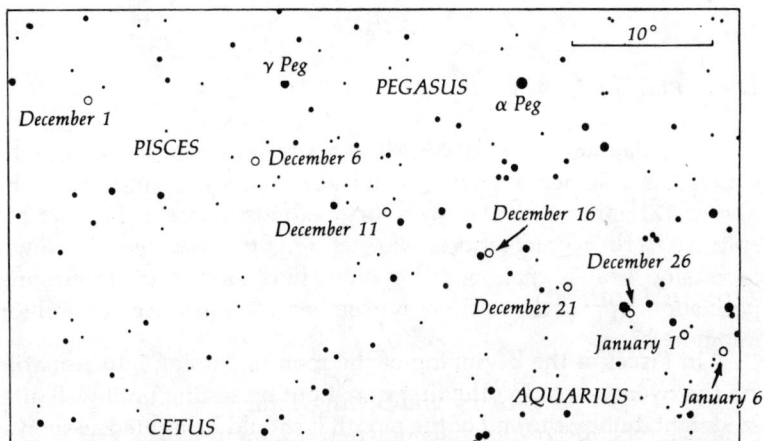

the right-hand part of the W. Extend it for the same distance and you will come to the bright star Alpha (α) Andromedae. This is the upper left-hand corner of the Great Square of Pegasus, below which the comet's route passes.

4. Now move down about 15° (remember that the outstretched fingers at arm's length measure about 20°), and you will come to Gamma (γ) Pegasi, with Alpha (α) Pegasi about 18° toward the right. These two stars are shown on the map.

The key chart shows the approximate positions of the guide stars for an observer somewhere in the middle of the "standard latitude" band, at an "average" time during the month in question. The same applies to other key charts in this chapter. They cannot be precise, but they should permit ready identification of the map region.

The Comet's Appearance

Observers in lower latitudes, in particular, may be able to make it out with the naked eye as a hazy "star" by the end of the month. Otherwise, binoculars will be necessary throughout December. We are observing the comet sideways-on during this month, so conditions are favorable for seeing any short tail that may have formed.

DECEMBER
1985

Earth and the Comet

The diagram shows that Earth is actually moving *away* from the comet, the distance increasing from 94½ million kilometers on the first to 171 million on the thirty-first—almost twice as far! Yet despite this, the comet appears to get brighter, because the Sun's increasing heat is making it far more luminous. The closest preperihelion approach was on November 27 (distance 93 million kilometers).

Rising and Setting Times
Boldface type means this phenomenon occurs in a dark sky.

Latitude 35° North

Date	Comet Rises	Comet South	Comet Sets	Twilight Ends	Altitude
Dec. 1	13.50	**20.30**	**03.10**	18.19	52°
6	12.50	**19.20**	**01.50**	18.19	59°
11	12.10	**18.20**	**00.40**	18.19	60°
16	11.30	17.40	**23.40**	18.21	56°
21	10.50	16.50	**23.00**	18.23	49°
26	10.20	16.20	**22.10**	18.26	44°
31	09.50	15.40	**21.40**	18.29	36°

Latitude 50° North

Date	Comet Rises	Comet South	Comet Sets	Twilight Ends	Altitude
Dec. 1	13.20	**20.30**	**03.40**	17.58	41°
6	12.30	**19.20**	**02.10**	17.57	45°
11	11.50	**18.20**	**00.50**	17.57	45°
16	11.20	17.40	**23.50**	17.58	42°
21	10.50	16.50	**23.00**	18.00	38°
26	10.20	16.20	**22.10**	18.02	35°
31	09.50	15.40	**21.30**	18.07	29°

Date	Time	Instrument	Appearance of Comet and Condensation	Appearance of Tail	Other Notes

J A N U A R Y

General Location

The comet remains in Aquarius throughout the month. This region is now moving into the evening twilight zone, and by the end of the month the comet will have been lost from view, being only 10° away from the Sun in the sky.

To Find the Map Region

1. The best "key star" is Epsilon (ε) Pegasi, shown at the top of the map. Find the Great Square of Pegasus, familiar from last month's watchings, and use the key chart to find Epsilon by extending the line from Gamma Pegasi through Alpha Pegasi for a little more than the same distance, and bending it slightly downward. Epsilon is about the same brightness as Alpha, although it may appear dimmer because of its lower altitude.
2. Then locate Alpha (α) Aquarii by "turning right" for somewhat more than 90° and heading for the horizon. This star, and its companion Beta (β), some 10° away and near the center of the map region, will serve as excellent pointers for the comet's path during January.

The Comet's Appearance

Due to the brightening surrounding sky, the comet may prove to be little more conspicuous than at the end of December, particu-

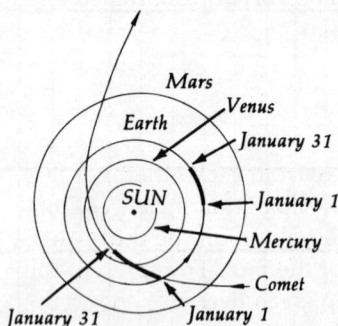

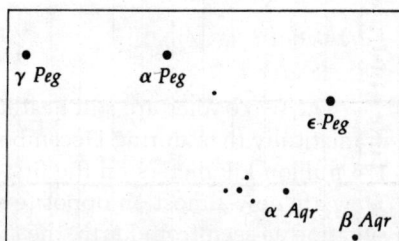

1 9 8 6

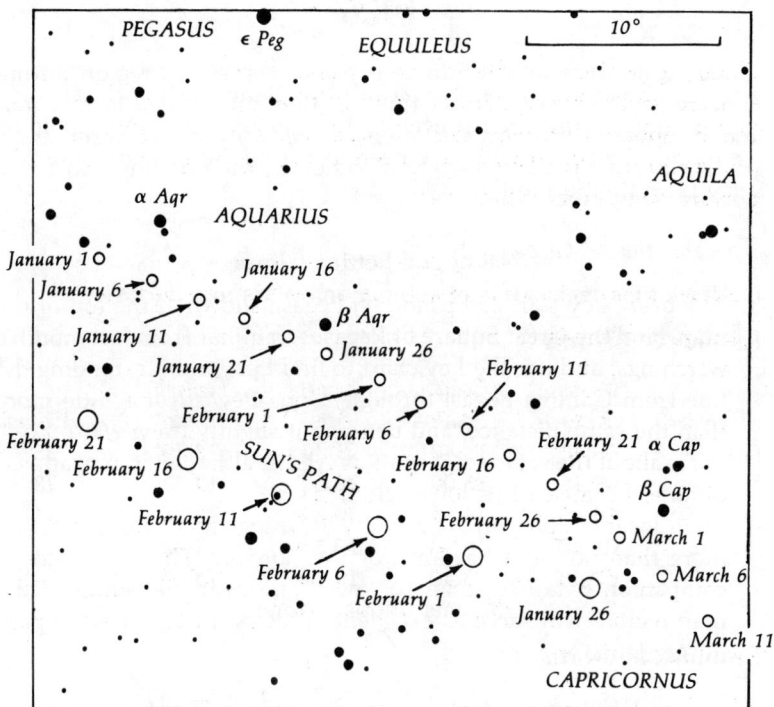

larly from higher latitudes. Binoculars will certainly help in locating it. Noticeable tail growth will have begun, but we now have an almost "nose-on" view, so that the tail will be very foreshortened and not easily seen.

Earth and the Comet

The two bodies are still heading away from each other, but less dramatically than during December; their separation increases from 174 million kilometers on the first to 234 million on the thirty-first. They are now almost on opposite sides of the Sun. To represent the situation as seen from Earth, the star map given here also shows the

JANUARY
1986

changing position of the Sun as it passes across the region during January and February. Try to think of the Sun as fixed in position, and its apparent leftward shift being caused by you, the observer, as you are carried in a left-to-right direction by the whirling Earth as it sweeps along in its orbit.

Rising and Setting Times
Boldface type means this phenomenon occurs in a dark sky.

Latitude 35° North

Date	Comet Rises	Comet South	Comet Sets	Twilight Ends	Altitude
Jan. 1	09.40	15.40	**21.30**	18.29	35°
6	09.15	15.00	**20.50**	18.33	28°
11	08.50	14.30	**20.20**	18.38	21°
16	08.20	14.10	**19.50**	18.42	14°
21	07.50	13.40	**19.20**	18.45	7°

(Comet unobservable)

Latitude 50° North

Date	Comet Rises	Comet South	Comet Sets	Twilight Ends	Altitude
Jan. 1	09.50	15.40	**21.20**	18.07	28°
6	09.20	15.00	**20.50**	18.12	23°
11	09.00	14.30	**20.10**	18.17	18°
16	08.30	14.10	**19.40**	18.23	12°
21	08.10	13.40	**19.10**	18.30	6°

(Comet unobservable)

Date	Time	Instrument	Appearance of Comet and Condensation	Appearance of Tail	Other Notes

F E B R U A R Y

General Location

The comet is in Aquarius until the twenty-fourth, when it passes into Capricornus. For most of the month it will be too near the Sun to be seen (perihelion passage occurs on the ninth), but observers in lower latitudes can try to recover it in the eastern sky, just before dawn, in the last few days of the month. It is unlikely to be seen from the 50° north latitude station until toward the end of April.

To Find the Map Region

1. The best guide stars are Alpha (α) and Beta (β) Capricorni.
2. Find the Pole Star, and lie on your back with your head to the north.
3. Referring to the small circular chart, start at the overhead point and shift your gaze toward the east (left) until you come to a very bright blue-white star. This is Alpha (α) Lyrae, commonly known as Vega.

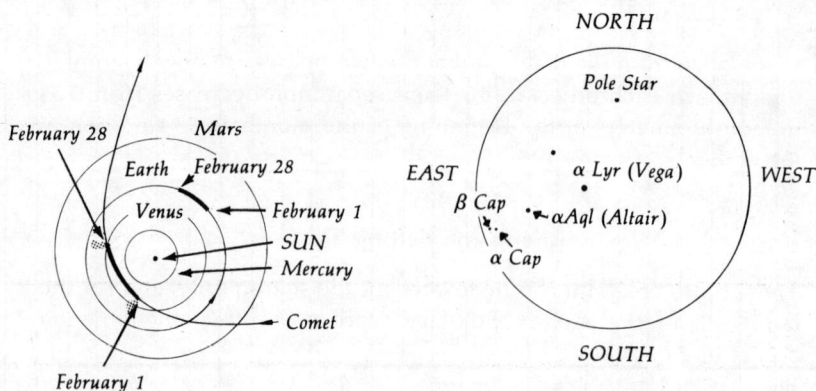

1 9 8 6

4. Having identified Vega, let your gaze descend about 30° toward the horizon. You will find Alpha (α) Aquilae, or Altair, set between two dimmer stars a couple of degrees above and below. It lies in a bright region of the Milky Way.

5. Now continue the line from Vega through Altair for about three quarters of the distance. You will come to Alpha and Beta Capricorni, very near the horizon. Alpha is distinctive, consisting of two stars very close together (binoculars will reveal them, if you are in doubt). Once you have located the Capricornus stars, you can judge the position of the comet's head from the map.

The Comet's Appearance

It will be such a difficult object in the dawn sky that you should not expect to see more than a hazy speck. By now, the tail is likely to be some 20 million kilometers long—but it is still directed somewhat away from Earth, and the twilight will hide it. The comet will be a binocular rather than a naked-eye object.

Earth and the Comet

Notice from the diagram that the two bodies are now beginning to approach each other again. Their separation decreases from 234 million kilometers at the beginning of the month to 194 million at the end.

Rising and Setting Times

These times are valid for the lower-latitude station only, and for the last week of the month. Boldface type means this phenomenon occurs in a dark sky.

FEBRUARY
1986

Latitude 35° North

Date	Comet Rises	Comet South	Comet Sets	Twilight Begins	Altitude
Feb. 21	05.20	10.40	16.00	05.16	(−1°)
26	**05.00**	10.10	15.30	05.11	3°
28	**04.40**	10.00	15.10	05.08	4°

Date	Time	Instrument	Appearance of Comet and Condensation	Appearance of Tail	Other Notes

M A R C H

General Location

During this month, the comet moves from Capricornus, through Sagittarius, and into Corona Australis. Since this motion is taking it southward, it will become more and more difficult to see from northern latitudes, and the head will not rise at all, before dawn breaks, as seen from a latitude of 50° north. The lower-latitude observer will have a reasonable view throughout the month, although the comet's altitude, at dawn, will never exceed 15°.

To Find the Map Region

Since the tail is now becoming much more conspicuous, it may well be possible to locate the comet at a glance—although, toward the end of the month, the tail will lie across the Milky Way and will therefore lose contrast with the sky. The stars Alpha and Beta Capricorni can still be used to identify the comet's position during the first part of the month, but you may, later, like to use the "Teapot" of Sagittarius, shown at the lower right-hand corner of the map.

1. Find the Pole Star and Vega, as described in the February notes.
2. Referring to the key chart, imagine a line drawn across from the Pole Star, through Vega, and then carried on for a little more than the same distance. This will bring you to the conspicuous group

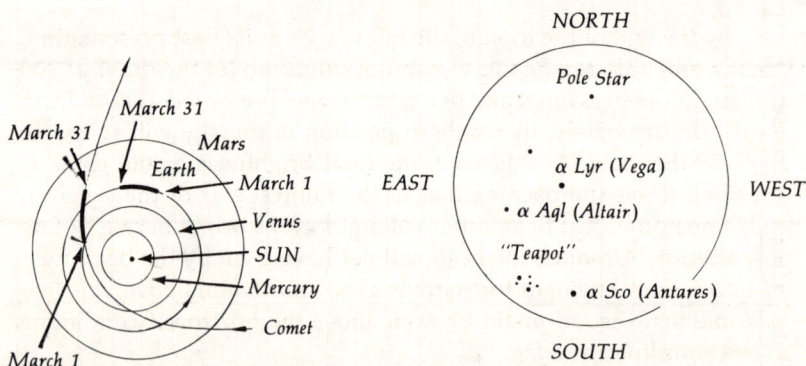

1 9 8 6

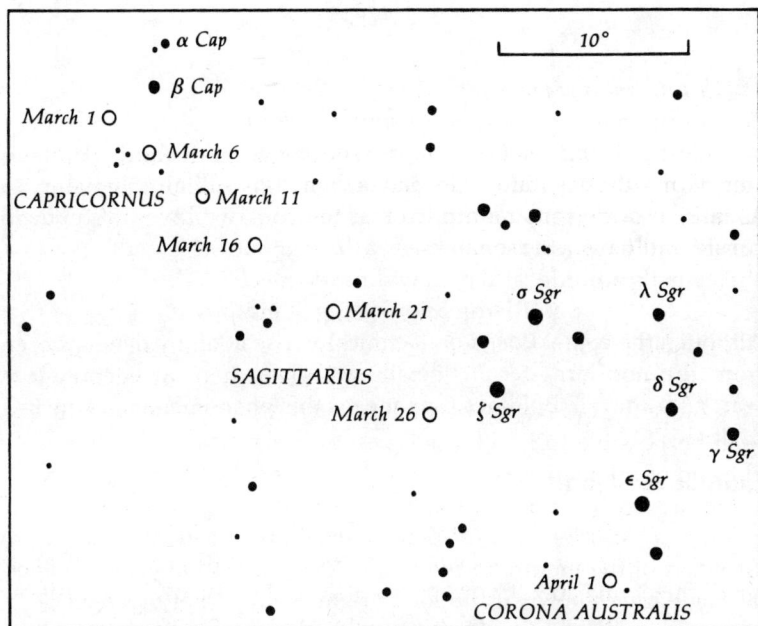

of stars made up of Zeta (ʓ), Sigma (σ), Lambda (λ), Delta (δ), Gamma (γ), and Epsilon (ε) Sagittarii: the "Teapot."

The Comet's Appearance

By the end of the month, the tail will be at its best presentation, having probably reached its maximum extension (of the order of 100 million kilometers long) and being fairly well presented as seen from Earth. Unfortunately, its southern position in the sky will seriously degrade the view. At this time, the total brightness of the coma is expected to be approaching that of the faintest star of the seven in the Dipper, or Great Bear; but it will not look as bright because of its low altitude. Although the head will not have risen by the beginning of astronomical twilight for stations at 50° north and above, it is just possible that the tail might be seen above the horizon if conditions are extraordinarily clear.

MARCH
1986

Earth and the Comet

Earth's orbital motion is now carrying it towards Halley's comet, and the separation decreases from 190 million kilometers to 82 million during the month, so that the coma will seem to expand considerably over these weeks.

Rising and Setting Times

Although the comet does not, technically, rise at all by dawn as seen from the northern station, details are given here to encourage a search for the tail. Boldface type means this phenomenon occurs in a dark sky.

Latitude 35° North

Date	Comet Rises	Comet South	Comet Sets	Twilight Begins	Altitude
Mar. 1	**04.40**	09.50	15.10	05.07	5°
6	**04.20**	09.30	14.30	05.01	8°
11	**04.00**	09.00	14.00	04.54	10°
16	**03.30**	08.20	13.10	04.47	12°
21	**03.10**	07.50	12.30	04.39	14°
26	**02.40**	07.00	11.20	04.32	15°
31	**02.10**	06.00	09.50	04.24	14°

Latitude 50° North

Date	Comet Rises	Comet South	Comet Sets	Twilight Begins	Altitude
Mar. 1	05.10	09.50	14.30	04.58	(−3°)
6	05.00	09.30	13.50	04.48	(−2°)
11	04.40	09.00	13.10	04.37	(−1°)
16	04.30	08.20	12.20	04.25	(−1°)
21	**04.10**	07.50	11.20	04.14	0°
26	04.10	07.00	10.00	04.02	(−1°)
31	04.20	06.00	07.40	03.49	(−3°)

Date	Time	Instrument	Appearance of Comet and Condensation	Appearance of Tail	Other Notes

A P R I L

General Location

During this month, Halley's comet moves across a very large angle of sky—from Corona Australis, through Scorpius and Centaurus, crossing Hydra, and ending in the little constellation of Crater. This long journey of some 120° takes it from the "morning sky"—in other words, a position best seen toward dawn—into the "evening sky," when it is at its highest angle above the horizon before midnight. Its southern dip reaches a maximum on April 10, after which it begins to move northward again. It will be poorly placed even for the lower-latitude observers around this time, and watchers from 50° north have little chance of a sighting until the last week or ten days of the month.

To Find the Map Regions

To accommodate the comet's extensive passage, it has been necessary to use two maps.

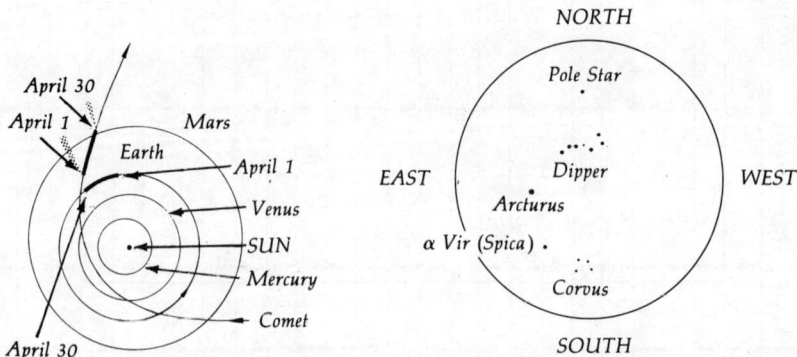

1 9 8 6

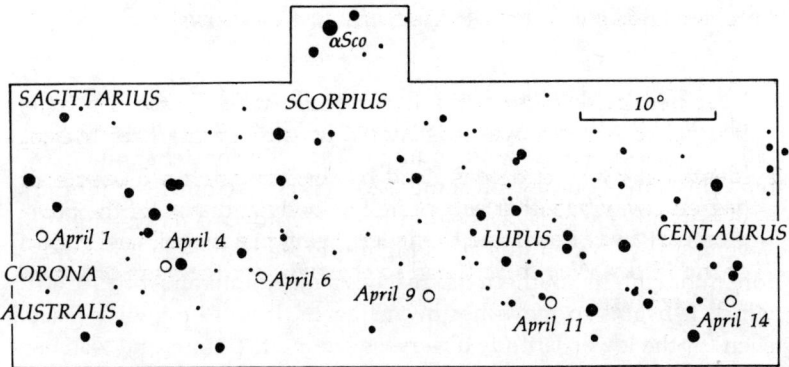

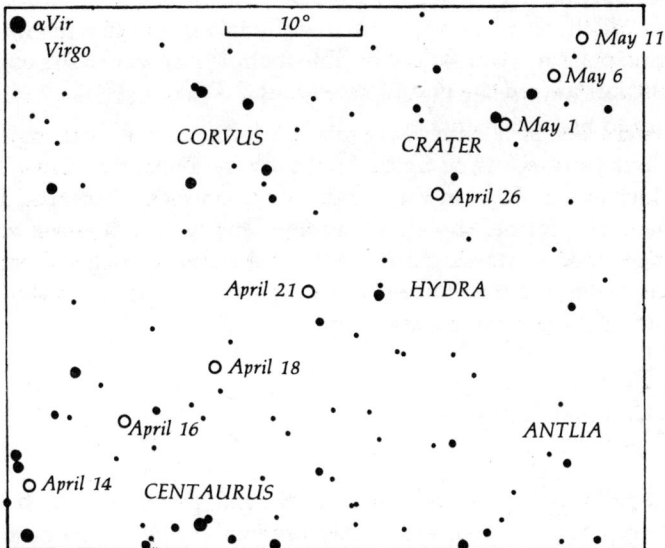

A P R I L

To Locate Map 1 (Corona Australis to Centaurus)

1. Your best guide to this area is the bright reddish star Alpha (α) Scorpii, or Antares, which is shown on the key chart for March.
2. Identify it by its rich color, and by the fainter stars a couple of degrees away on either side of it. The body and sting of the Scorpion curl away down and to the left, lying in a magnificent region of the Milky Way. Below and to the right are the stars of Lupus and Centaurus.

To Locate Map 2 (Centaurus to Crater)

1. Probably the most useful starting point is the small but distinctive constellation of Corvus, the Crow. By taking sights from its four principal stars, you should be able to find your way to the comet's position toward the end of the month.
2. Corvus lies in a rather bare patch of sky, to the west (right) of Alpha (α) Virginis, or Spica. Start with the Dipper, or Great Bear, which is very high in the northern sky on April evenings. Trace the curve formed by the "handle," and you will come to the bright reddish star Arcturus, in the constellation Boötes. Continue this curve, and it will bring you to Spica, in the southern sky. The route is shown on the key chart.

The Comet's Appearance

April sees a rapid shortening of the tail, partly because it really is likely to get shorter, and partly because it appears to pass "behind" the coma in mid-month as Earth passes between it and the Sun. There will also be a general fading of its brightness, and despite its improved altitude by the end of the month, it will probably have

1 9 8 6

sunk to below naked-eye visibility—remaining so for another three quarters of a century!

Earth and the Comet

The diagram shows how our planet's movement makes the tail appear to swing from the right side of the comet at the beginning of the month to the left side at the end. Closest approach is on April 11, when the two bodies are 63 million kilometers apart. By the end of the month, this will have increased to 116 million kilometers.

Rising and Setting Times

During this month the comet is well away from the Sun in the sky, and twilight conditions are of little importance. The "altitude" column therefore refers to the height of the comet above the horizon when it is due south in the sky. Boldface type means this phenomenon occurs in a dark sky.

Latitude 35° North

Date	Comet Rises	Comet South	Comet Sets	Altitude When South
Apr. 1	**02.00**	05.50	09.30	14°*
6	**01.10**	**04.10**	07.10	10°
11	**23.00**	**01.40**	04.30	8°
16	20.00	**23.30**	**03.00**	15°
21	17.40	**22.00**	**02.20**	24°
26	16.10	**21.00**	**01.50**	31°
30	15.20	**20.30**	**01.30**	36°

* On this date, astronomical twilight commences at 04.24, before the comet is due south. The altitude value is calculated for the onset of twilight.

APRIL
1986

Latitude 50° North

Date	Comet Rises	Comet South	Comet Sets	Altitude When South
Apr. 16	—	23.30	—	(0°)
21	19.00	22.00	00.50	9°
26	17.10	21.00	01.00	16°
30	16.10	20.30	00.50	19°†

† On this date, astronomical twilight ends at 21.36, after the comet is due south. The altitude value is calculated for the end of twilight. On the twenty-sixth, the comet is south a few minutes before twilight ends, but the difference is negligible.

April 1986

Date	Time	Instrument	Appearance of Comet and Condensation	Appearance of Tail	Other Notes

M A Y

General Location

The comet moves across only about 15° of sky this month, passing from Crater, through a corner of Hydra, into the obscure constellation of Sextans, in the southwestern sky after sunset.

To Find the Map Region

On the first, the comet is conveniently located near Alpha (α) Crateris. Use the key chart to locate Crater from the neighboring group of Corvus, which you had to identify in April. Note that the faintest stars shown here are somewhat dimmer than those on the previous maps, since the comet will be fainter and may need more signposts to permit identification.

The Comet's Appearance

This month will probably see the comet lost from view for observers at 50° north, since the summer twilight will linger more and

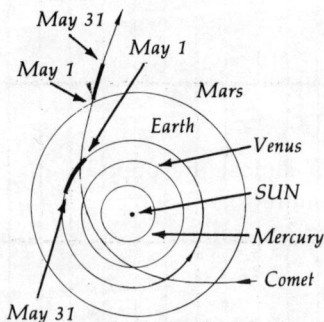

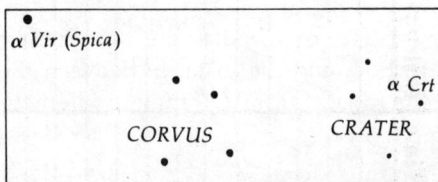

1 9 8 6

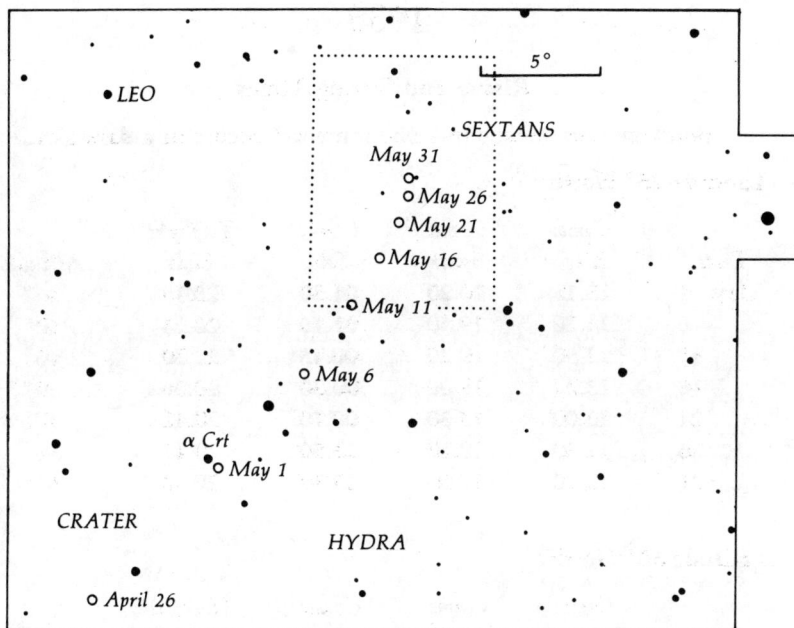

more obtrusively as May wears on. Even in the favorable skies of southern latitudes it will be a difficult binocular object by the end of the month.

Earth and the Comet

The two bodies are now moving in almost the opposite direction, and the distance between them more than doubles during the month, from 120 million kilometers to 266 million.

MAY
1986

Rising and Setting Times

Boldface type means this phenomenon occurs in a dark sky.

Latitude 35° North

Date	Comet Rises	Comet South	Comet Sets	Twilight Ends	Altitude
May 1	15.15	**20.20**	**01.30**	20.18	36°
6	14.20	19.50	**01.10**	20.24	40°
11	13.50	19.20	**00.45**	20.30	40°
16	13.20	19.00	**00.30**	20.36	39°
21	13.00	18.30	**00.10**	20.42	37°
26	12.30	18.10	**23.50**	20.47	34°
31	12.10	17.50	**23.30**	20.52	31°

Latitude 50° North

Date	Comet Rises	Comet South	Comet Sets	Twilight Ends	Altitude
May 1	15.50	20.20	**00.50**	21.36	20°
6	15.00	19.50	**00.40**	21.52	20°
11	14.20	19.20	**00.20**	22.10	18°
16	13.40	19.00	**00.10**	22.27	14°
21	13.10	18.30	23.50	22.45	11°
26	12.50	18.10	23.30	(23.00)	(6°)*
31	12.20	17.50	23.20	(23.00)	(4°)*

* As seen from this station, twilight lasts all night from the end of May until the middle of July, but the sky is fairly dark from about 23.00 until 01.00 local time. However, it seems very unlikely that any sightings of the comet will be possible.

Date	Time	Instrument	Appearance of Comet and Condensation	Appearance of Tail	Other Notes

J U N E

General Location

The comet is now heading almost directly away from Earth, and will appear to move across only a few degrees of sky, remaining in the constellation Sextans. It will already have been lost from view by observers at 50° latitude, and the region is rapidly sinking in the southwest for those at 35° latitude.

To Find the Map Region

The large-scale map given here corresponds to the rectangle shown on the map for May. The faintest stars shown here are visible only with binoculars, even when the sky is completely dark and the region is high in the sky.

The Comet's Appearance

By the end of the month, the comet will probably be beyond the reach of ordinary binoculars, and will be an exclusively "telescopic" object, resembling a dim hazy star.

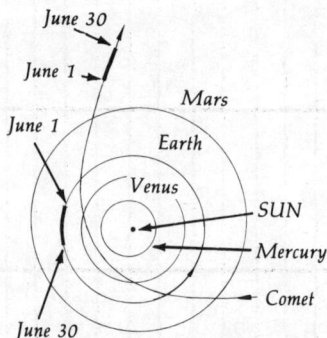

1 9 8 6

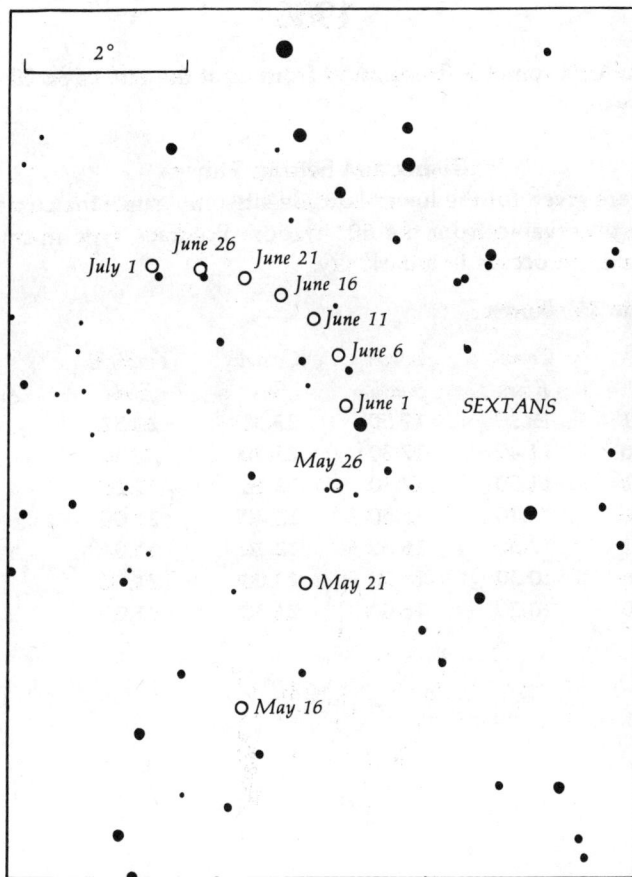

Earth and the Comet

The diagram shows that the two bodies are moving directly away from each other, the distance increasing from 266 million kilometers to 406 million during the course of the month. This means

JUNE
1986

that Halley's comet is flying away from us at the rate of 54 *kilometers per second!*

Rising and Setting Times

These are given for the lower-latitude site only, since the comet will not be observable from the 50° station. Boldface type means this phenomenon occurs in a dark sky.

Latitude 35° North

Date	Comet Rises	Comet South	Comet Sets	Twilight Ends	Altitude
June 1	12.10	17.50	**23.30**	20.52	30°
6	11.40	17.30	**23.10**	20.56	26°
11	11.20	17.10	**22.50**	21.00	23°
16	11.10	16.50	**22.40**	21.02	19°
21	10.50	16.30	**22.20**	21.04	15°
26	10.30	16.20	**22.00**	21.05	12°
30	10.20	16.00	**21.50**	21.05	9°

Date	Time	Instrument	Appearance of Comet and Condensation	Appearance of Tail	Other Notes

GEORGE PHILIP

Also from George Philip

Astronomy with a Small Telescope
James Muirden
Star Maps by Wil Tirion

The complete handbook for the beginner written by an
acknowledged expert

✻ How to choose and set up a telescope

✻ What to look for in the night sky and how to find it

✻ What you can see with your eyes alone, with binoculars, or
with a small telescope

✻ Month-by-month star charts for both the northern and the
southern hemisphere

✻ Detailed charts of the most interesting constellations

✻ Background information on everything from comets,
meteors and sunspots to quasars, nebulae and spiral galaxies

✻ How to photograph the stars

£9.95 net (Published November 1985)